CHANGING
GEOGRAPHY

SERIES EDITOR: **JANET SPEAKE**

Glaciers and glacial landscapes

PETER G KNIGHT

Geographical
Association

ACKNOWLEDGEMENTS

Peter Knight gratefully acknowledges the help and support of teachers and colleagues at King Edward VI School, Edgbaston, Birmingham, and in the Physical Geography teaching team at Keele University, for instruction, advice and assistance over many years that have contributed to the production of this book. Special thanks are also due to Zena Furby, Dr Debbie Knight and Professor John Bale for their encouragement and assistance.

AUTHOR: Peter G. Knight is Senior Lecturer in Physical Geography, Keele University.

ISBN 1 84377 097 0
First published 2005
Impression number 10 9 8 7 6 5 4 3 2 1
Year 2008 2007 2006 2005

Published by the Geographical Association, 160 Solly Street, Sheffield S1 4BF. The Geographical Association is a registered charity: no 313129.

The Publications Officer of the GA would be happy to hear from other potential authors who have ideas for geography books. You may contact the Officer via the GA at the address above.

Edited by Tony Williams
Designed by Arkima Ltd, Dewsbury
Printed and bound in China through Colorcraft Ltd., Hong Kong

CONTENTS

EDITOR'S PREFACE

The books in the *Changing Geography* series seek to alert students in schools and colleges to current developments in university geography. The series also aims to close the gap between school and university geography. This is not a knee-jerk response – that school and college geography should be necessarily a watered-down version of higher education approaches – but a deeper recognition that students in post-16 education should be exposed to the ideas currently being taught and researched in universities. Many such ideas are of interest to young people and relevant to their lives (and school examinations).

The series introduces post-16 students to concepts and ideas that tend to be excluded from conventional school texts. Written in language which is readily accessible, illustrated with contemporary case studies, including numerous suggestions for discussion, projects and fieldwork, and lavishly illustrated, the books in this series put post-16 geography in the realm of modern geographical thinking.

The study of glaciers is increasingly important in the context of environmental change, and recent developments in the subject have seriously challenged traditional approaches. *Glaciers and glacial landscapes* reviews our basic understanding of glaciers within the global system and identifies ways in which that understanding is changing in response to new scientific research.

Janet Speake
January 2005

INTRODUCTION

The study of glaciers is important not only for understanding how landscapes are created, but also for understanding global issues such as climate change and sea-level rise. This book examines the characteristics of glaciers, explores how they shape the landscape, and explains their role in the unfolding drama of global environmental change. It also considers the impact of glaciers on human activity, and the potential impact of humans on the future growth or melting of the world's ice.

Research on glaciers is carried out by a wide range of scientists including physicists, chemists, climatologists, oceanographers and mathematicians. Their work involves field observation, laboratory experiments, satellite remote sensing and numerical modelling. This book looks particularly at the work of glaciologists (who study ice and snow) and geomorphologists (who study landforms). Their research has progressed so rapidly that school curricula and textbooks struggle to keep up with new developments in the subject. Recent discoveries have led to major reassessments of fundamental questions such as how glaciers move, how they erode rock and deposit sediment, and how they respond to global warming. These new discoveries challenge traditional views of glaciers and glaciated environments.

Because of their impact on hydrological and climatic systems, glaciers have a global impact. Tropical ocean currents are affected by glaciers in the Arctic, and Britain's coastline could be devastated by ice melting in Antarctica. Glaciers hold more than 75% of the world's fresh water, and provide irrigation for some of the planet's most densely populated areas. Ten per cent of the Earth's land surface is covered by glaciers. Twice this area has been exposed by retreating glaciers in the last 20,000 years. These glaciated landscapes include most of Great Britain as well as much of northern Europe and North America.

This book will help you to learn about the importance of glaciers, and to explore some of the new questions that science is asking about glaciers. To help you in your learning, Information Boxes and Case Studies are used to provide extra detail about specific issues, and Activity Boxes offer practical activities to help you focus your thinking and develop your understanding of important ideas.

CHAPTER 1

GLACIERS AND GLACIATION

Glaciers are defined as bodies of ice that form by the accumulation of snow on the surface of the ground and that are thick enough to deform and 'flow' under their own weight. It is important to realise that glaciers do not start life as frozen rivers or lakes, but begin as thick masses of snow. The snow accumulates year after year until it is so thick that the lower layers are compressed into ice that can spread outwards, or flow downhill, rather like a pile of very gooey porridge. Other types of ice in the landscape, such as lake ice or sea ice formed by freezing of a water surface, cannot be called glaciers. So, for example, there is no glacier at the North Pole where the surface of the ocean is frozen, but there is a glacier at the South Pole, where layer upon layer of snow has accumulated on the ground in Antarctica to build an ice sheet 4km thick.

How do glaciers form?

Glaciers begin to form when the snow that falls each winter does not all melt away the following summer. The snow then accumulates layer upon layer, year after year. As the snow becomes thicker, the lower layers are compressed under the weight of the snow above and gradually turn into ice. As the snow continues to accumulate on the top surface, and the ice becomes thicker and thicker, it begins to deform under its own weight. As the ice deforms it begins slowly to move, or flow, and can then be called a glacier (see Information Box 1 and Activity Box 1).

Mass balance, accumulation and ablation

The formation of glaciers depends on the balance between: the input of ice from sources such as snowfall, freezing of water or avalanches

---

Information Box 1: Transformation of snow to ice

Snow is transformed to ice by compression, by recrystallisation, and in some environments by melting and refreezing. Fresh snowfall has a very open structure with lots of interconnecting air spaces. As snow is compressed under the weight of more snowfall, the structure is compressed, air spaces shrink and the density increases. Snow crystals grow together into larger ice crystals by processes known as annealing and sintering. When all the connections between airspaces are closed, and air exists only in isolated bubbles, the transformation to ice is complete (Figure 1). In cold, dry environments this process can take hundreds of years, but in warmer, wetter environments, where meltwater is produced and refreezes to make ice within the snow, the process can be completed within one or two years. The characteristics of snow, of glacier ice, and of the intermediate material called firn, are shown below.

Material	Density (kg m³)	Porosity
Snow	Low (10-550)	High
Firn	Medium (400-840)	Medium
Ice	High (830-917)	Low

Figure 1: A piece of glacier ice cut from the edge of the Greenland ice sheet. The bubbles in the ice are formed from air that was trapped as the snow turned to ice thousands of years ago. Photo: Peter G. Knight.

(accumulation); and the loss of ice by processes such as melting and the calving off of icebergs (ablation). This balance of accumulation and ablation is referred to as the *mass balance* of a glacier. Where there is more accumulation than ablation the balance is positive and the glacier is growing. Where there is more ablation than accumulation the balance is negative and the glacier is shrinking. Where the accumulation and ablation are equal the glacier is said to be in equilibrium. This can apply to the whole glacier or just to one part of the glacier. For example, most glaciers have an upper *accumulation area* where most of the snow falls and there is little melting, and a lower *ablation area*, dominated by melting or iceberg calving. The boundary between the two zones is known as the *equilibrium line*, which marks the level on the glacier where the annual accumulation exactly matches the annual ablation (see Case Study 1). The altitude of the equilibrium line (the ELA) varies with climate, and can be used to indicate the health of the glacier. The higher the ELA, the less material is fed into the glacier and the more material is lost, so a rising ELA is usually bad news for the future of a glacier (see Chapter 4).

What are glaciers made of?

Glaciers are made primarily of:
1. ice derived from the compression and recrystallisation of snow

---

Activity Box 1: Looking at ice crystals

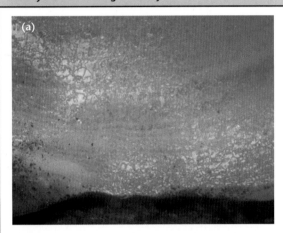 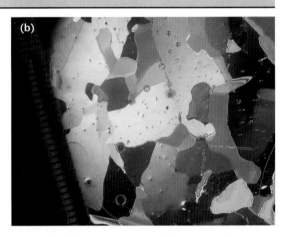

Figure 2: Ice crystals: (a) in a glacier showing its polycrystalline structure, and (b) polycrystalline ice viewed through a polarising filter. Photo: Peter G. Knight.

The crystals in the glacier ice shown in Figure 2a are clearly visible because slight melting has occurred at the boundaries between the crystals. However, the individual crystals of ice can usually be seen more clearly if a thin slice of ice is viewed through a light-polarising filter of the type used with cameras and polarising sunglasses (Figure 2b). For the best effect you should have one polarising lens behind the ice sample and one in front of it, but you will get some effect if you simply look at the ice through a single polarising filter. Because crystals that are orientated at slightly different angles transmit the polarised light in slightly different ways, different crystals appear as different colours.

- Wear thick gloves to protect your fingers, and safety goggles to protect your eyes.
- Take a slice about 1mm thick from a piece of ice. You can either cut the ice with a small hacksaw or melt it down to a thin slice in a frying pan over a low heat. Use tongs to hold and turn the ice.
- Place this 'thin section' of ice on a light-box, or hold it up to the light.
- Ideally, place one polarising filter behind the ice, and another in front of it. If you have only one filter, place it in front of the ice and view the ice through it.
- Twist the filter nearest to you until you achieve the best effect.
- Try looking at thin sections from: an ice cube from your freezer; a tightly squeezed snowball; a frozen pond surface; and other specific sources. Compare and describe the sizes and shapes of the crystals in the different types of ice.

2. rock debris from material falling onto the glacier's surface
3. rock debris eroded and entrained from the glacier bed
4. water derived from melting of snow and ice
5. air in the form of bubbles in the ice.

Different glaciers contain these elements in different amounts. For example, glaciers in warm locations contain more water than glaciers in colder locations. Ice in glaciers is created by two main processes: transformation of snow (see Information Box 1) and freezing of water. Freezing of water can occur in two situations within a glacier:

1. at the surface when water from snow melted by the sun refreezes to make layers of ice as it percolates downwards into the deeper snow pack
2. at the base when meltwater flowing beneath the ice freezes on as a layer of *basal* ice at the sole of the glacier.

The structure of ice

Glacier ice is a polycrystalline solid. This means that it is made up of many individual crystals joined together to make the ice mass. The ice crystals in glaciers are typically a few millimetres or centimetres in diameter, and fit closely together like a jigsaw puzzle (Figure 2b). The crystal boundaries are marked by veins just a few microns thick (a micron is one thousandth of a millimetre), and water can exist within these veins even when the temperature is below 0°C.

The structure of ice at the atomic scale is very important for the behaviour of glaciers. Ice is made of atoms of oxygen and hydrogen joined together as water molecules (H_2O) and fixed into a rigid crystal lattice. Within each crystal, molecules are arranged in roughly parallel sheets, almost like a deck of cards. This is important because it means that crystals can be deformed more easily in some directions than in others (see 'How glaciers move', Chapter 4, page 29).

Information Box 2: Different types of glacier

Ice sheet: a large glacier (larger than 50,000km²) that completely submerges the regional topography and forms a gently sloping dome of ice that can be several kilometres thick in the centre. The Greenland and Antarctic ice sheets are the only ones that exist at present, but others have existed during former glacial periods. The Laurentide ice sheet, which covered the northern part of North America during the last ice age, was larger than today's Antarctic ice sheet.

Ice stream: a linear zone of faster flowing ice within an ice sheet, often corresponding to a deep trough in the subglacial topography.

Ice cap: similar to an ice sheet but smaller than 50,000km². For example, a glacier engulfing Wales or Scotland would be called an ice cap rather than an ice sheet simply because of its smaller size. Many ice caps occupy upland areas, and do not extend far beyond the edge of their highland source. Examples include Vatnajökull, in southern Iceland, which is Europe's largest existing ice cap at about 8400km². The term mountain ice cap describes a small ice cap that covers an individual mountain summit.

Ice field: ice covering an upland area including mountain peaks and valleys but not thick enough to bury the topography completely.

Cirque glacier: a small glacier that occupies a topographic hollow on a mountainside and enlarges it into a deep basin. Also sometimes referred to as a *corrie* glacier. If the cirque glacier grows beyond the size of its hollow then ice can flow out to form a larger valley glacier. Glaciers too small to develop cirques are sometimes called *niche glaciers*.

Valley glacier: a glacier confined between valley walls and terminating in a narrow tongue. *Alpine valley glaciers* are supplied by ice flowing out of mountain cirques. Outlet valley glaciers are supplied by ice flowing out from the edge of an ice sheet, ice cap or ice field. When a valley glacier extends beyond the end of a mountain valley into a flatter area, the ice can spread outwards like a fan to form a *piedmont glacier*. Valley glaciers that terminate in the sea, such as those that flow into fjords, are sometimes referred to as *tidewater glaciers*.

Ice shelf: a large area of floating glacier ice extending from the coast where several glaciers extend into the sea and coalesce. The large Ross and Ronne ice shelves that extend from the Antarctic ice sheet cover areas the size of Spain and are up to 2000m thick.

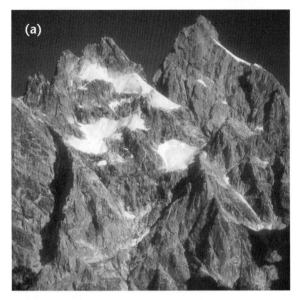

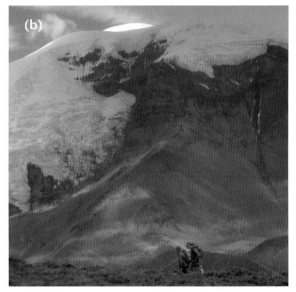

Figure 3: Glaciers come in different shapes and sizes: (a) tiny niche glaciers in the Grand Teton mountains in the USA; (b) the ice cap on the summit of Chimborazo, Ecuador; and (c) a small part of the margin of the Greenland ice sheet. Photos: Peter G. Knight.

Where do glaciers exist today?

Glaciers exist wherever the local climate allows more ice to accumulate each year than can be removed by ablation: in other words wherever the mass balance is positive. As climate changes, so does mass balance, and that is why glaciers grow and shrink, and appear and disappear, through time. Today, glaciers exist in a variety of locations all over the world (see Activity Box 2). There are large ice sheets covering Greenland and Antarctica, and in the polar regions there are many smaller glaciers and ice caps reaching down as low as sea level. In mid latitudes,

for example in the European Alps, few glaciers reach below 2000m. At the Equator, glaciers only occur above about 5000m, but even on the Equator, snow can accumulate to form glaciers at high altitudes. In Ecuador, for example, volcanoes such as Cayambe and Cotopaxi, which lie on the Equator but reach heights of around 5800m above sea level, have substantial ice caps.

There are many different types of glacier. Figure 3 shows how glaciers come in many different shapes and sizes, and Information Box 2 lists some of the most important types of glacier.

Activity Box 2: Independent research

1. Use an atlas, or search the internet, to find out more about these examples of different types of glacier. For each example, try to find out its latitude, the altitude above sea level of its lowest point, its surface area and its maximum thickness. Copy the table below, and insert your additional information into the fourth column.

Type of glacier	Examples	Location	Additional information
Ice sheet	The Greenland ice sheet	Greenland	
	The Antarctic ice sheet	Antarctica	
Ice cap	Vatnajökull	Iceland	
Ice field	Patagonian Ice field	Chile	
Mountain ice cap	Kilimanjaro	Tanzania	
	Cotopaxi	Ecuador	
Valley Glacier	Aletsch Glacier	Switzerland	
	Athabasca Glacier	Canada	
Piedmont Glacier	Malaspina Glacier	Alaska	
Ice shelf	Ross Ice Shelf	Antarctica	

2. Mark the location of each of these glaciers on a world map. For each glacier write its name, and the latitude and elevation above sea level of its lowest point.
3. Draw a graph of latitude against altitude to show how glaciers at higher latitudes reach down to lower altitudes than those nearer the Equator.
4. Whenever you hear or read about a new glacier, add it to your table and your map.
5. From the information that you collect, make a map that shows the global distribution of glaciers at the present. Compare your map with Figure 5, which shows the distribution at the last glacial maximum.

Case Study 1: The Greenland Ice Sheet

The Greenland ice sheet is one of the world's two remaining ice sheets. It covers an area of 1.7 million km^2 and contains more than 2.5 million km^3 of ice. If all this ice melted suddenly it would raise global sea level by approximately 6m. In the centre of the ice sheet the ice is more than 3km thick, and the weight of the ice depresses the Earth's crust by about 1km below the level that it would be at if it were not loaded with ice. Because of this depression, the land surface beneath central Greenland now lies below sea level (Figure 4).

Greenland's ice survives only because it creates its own local climate that supports it. The height of the ice surface above sea level means that temperatures are very low, precipitation falls as snow, and the mass balance is positive. If the ice sheet disappeared, the height of the exposed land surface would be so low that surface temperatures would be warmer, more precipitation would fall as rain rather than snow, the mass balance would be negative and the glacier would not grow back under the present climate.

In response to recent climate change Greenland's ice sheet appears to be melting and getting thinner around the edges, but getting thicker in the centre because the amount of snowfall it receives in that area increases with global warming. We really do not know what will happen to the ice sheet in the future, and this is a major area of current scientific research.

The estimates in Table 1 suggest that Greenland's ice sheet has a slightly negative balance and is therefore shrinking. However, the amount of uncertainty in measurements of accumulation and ablation is such that we don't actually know for sure whether the mass balance total is positive or negative. We will look at this in more detail in Chapter 4.

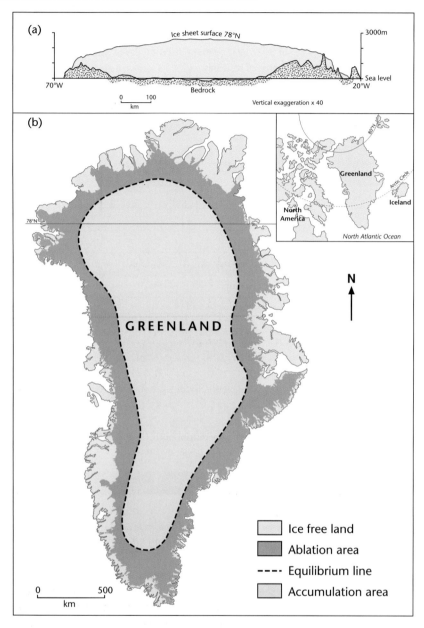

Table 1: Accumulation and ablation of the Greenland ice sheet.

Activity	Change per year (water km³)
Accumulation (mostly from snowfall)	+520
Ablation by melting	−290
Ablation by calving (icebergs)	−200
Ablation by sublimation (evaporation from the ice)	−60
Mass balance	−40

Note: figures are given in cubic kilometres of water per year, and are estimates based on many different pieces of research.

Figure 4: The Greenland ice sheet: (a) cross-section; and (b) ice sheet accumulation area, ablation area and equilibrium line.

Where have glaciers been in the past, and when?

In the past, glaciers have been much more extensive than they are today. Throughout the Earth's history, glaciers have expanded and contracted repeatedly. Some geologists believe that there have been periods in the distant past when the whole planet has been completely blanketed in ice. This idea is known as the 'Snowball Earth' theory, and although the theory is very controversial, it is certain that there have been repeated glacial periods or *Ice Ages* when glaciers did expand to cover large areas of the planet, as well as warmer periods or *interglacials* when glaciers disappeared completely. The Earth has been in a glacial period for about the last 2 million years, during which time glaciers have expanded and contracted repeatedly. The last glacial maximum occurred about 20,000 years ago. Although we are still in a long-term glacial period, the world's glaciers have retreated a long way since then (Figure 5).

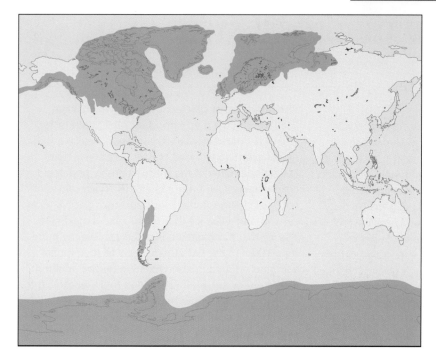

Figure 5: The extent of the major ice sheets at the last glacial maximum.

Activity Box 3: The last British ice sheet

Using textbooks, an encyclopedia or internet sources, try to answer the following questions:

■ At what periods in history have ice sheets expanded across Britain?

■ When did the most recent glaciation reach its maximum?

■ When did the last glaciers in Britain finally disappear?

■ What name is given to the brief re-advance of glaciers at the end of the last ice age?

Annotate a map of Britain to show:

■ The outline of the last British ice sheet at its maximum extent

■ Areas that experienced mountain glaciation after the main ice sheet had disappeared.

Keep the map you have made for future reference.

Conclusion

Glaciers can occur when more snow accumulates than melts each year. Today, glaciers exist only at high latitudes or high altitudes, whereas in the past they have been much more extensive. Periods of glaciation have occurred throughout geological history, and their effects are clearly visible in the landscapes of formerly glaciated areas such as Great Britain. Glaciers come in all shapes and sizes ranging from continental ice sheets to tiny mountain glaciers. Nevertheless, these different glaciers behave in similar ways. An understanding of glaciers will help us to understand the landscapes around us, the environments of the past, and the possible future consequences of environmental change, which is the focus of Chapter 2.

GLACIERS AND THE GLOBAL SYSTEM

How much ice?

Planet Earth can be looked at as a complex physical system in which components such as climate, sea level, ocean currents and the hydrological cycle are all connected. To get a feeling for how important glaciers are in the global system, it is useful to think about how much ice there is in the world's glaciers (Activity Box 4). Glaciers today contain about 26 million km³ of ice, and because so much of the world's water is locked up in glaciers, they form an important part of the global hydrological cycle. Because glaciers change in size through time, glaciation can cause major changes in other parts of the hydrological system, including sea level.

During the last ice age (at the last glacial maximum) there was more than twice as much ice as today, and therefore less water available to the oceans. Table 2 indicates the amounts of ice in glaciers now and during the last glacial maximum, and how much difference that much ice would make to sea level if it melted.

The influence of glaciers on sea-level change

Sea-level change is an important topic, not just for scientists but for all of us. It has been predicted that global warming will lead to rises in sea level that could flood many of the world's densely populated coastal areas. Glaciers are one of the main causes of sea-level change. Table 2 indicates that glaciers today contain much less ice than at the glacial maximum. The ice that has melted has raised sea level by more than 100m in the last 20,000 years or so. If global warming resulted in our remaining glaciers melting, even more water would be added to the oceans: the water produced would be sufficient to raise sea level by about 65m. Activity Box 5 investigates the effects of sea-level change.

With 65m of sea level locked up in glaciers, melting even a small proportion of the world's glaciers would create severe problems. However, the impact of global warming on glaciers is complex. Most scientists agree that global warming will quickly melt mountain glaciers in low latitudes (see Information

---

Activity box 4: Visualising the world's water

Before you start this activity, have a guess what proportions of the world's water are in glaciers, rivers and oceans. Write down your guess, and imagine what those proportions would look like side by side. For example, do you think there is twice as much water in rivers as in glaciers, or ten times as much in glaciers than in rivers? In fact glaciers today contain about 2% of all the Earth's water (including the oceans), but that's more than 75% of the planet's fresh water. This experiment will help you to visualise how much water is in each part of the hydrological system.

- Pour 1 litre of water into a measuring beaker – this represents all of the water on the planet.

- From this litre, pour off 28 ml into a second beaker. What is left in the first beaker represents the world's oceans, and your second beaker represents all the world's fresh water, including rivers, lakes, groundwater, glaciers and water in the atmosphere!

- From your small beaker of 'fresh water' pour off 6.5 ml into a test tube. The 21.5 ml left in your second beaker represents the water locked up in the world's glaciers. The 6.5 ml in your test tube is all of the world's remaining fresh water.

- From the 6.5 ml in your test tube draw out 0.5 ml with an eye-dropper. The 6 ml that you have left in your test tube represents groundwater, and the 0.5 ml in your dropper is all the water sitting on the Earth's surface as rivers and lakes and in the atmosphere.

- Try squeezing all the world's lakes and rivers out of the dropper into your hand. The tiny amount that you can't shake out of the dropper represents the amount of water in the world's rivers.

- Compare the volume of water in the world's rivers (0.0001%) with the amount of water in glaciers (your test tube, 2.15%) and the amount in the oceans (your first beaker, 97.2%).

Table 2: Volumes of ice in the ice sheets today and at the last glacial maximum, and the equivalents of those volumes expressed in terms of how much difference they would make to sea level if they were poured into, or taken from, the world's oceans

Ice sheet	Volume (million km^3)		Sea-level equivalent (m)	
	Today	(at glacial maximum)	Today	(at glacial maximum)
Antarctic	23.5	(26)	59	(66)
Greenland	2.6	(3.5)	6	(11)
Laurentide	0	(29.5)	0	(74)
Cordilleran	0	(3.6)	0	(9)
Scandinavian	0	(13.3)	0	(34)
Other	0.2	(1.1)	0.5	(3)
Total	26.2	(77)	65	(197)

Box 3). In areas where mean annual temperatures are close to the melting point, slight warming will lead to great increases in melting, so mountain glaciers are very susceptible to destruction by global warming. However, it is less certain how global warming will affect the size of the world's two ice sheets (Greenland and Antarctica). American Scientist Online features an article explaining how scientists disagree about the effect of global warming on ice sheets (see websites, page 55). Some scientists argue that warming will increase evaporation from the world's oceans, increase moisture content in the atmosphere, and therefore increase snowfall and glaciation. Warmer temperatures also mean more melting, but the mean annual temperature in Antarctica is well below the freezing point, so the overall effect of global warming may be to increase snowfall, to have little effect on melting, and therefore to cause net growth of the ice sheet. However, other scientists argue that increased melting will outweigh increased snowfall, and others argue that global warming will raise the temperature of the oceans so that the floating ice shelves around Antarctica will melt. The ice sheet may then become unstable, flow much more rapidly, and therefore shrink. This uncertainty is a serious problem in predicting future sea-level change, but as your maps from Activity Box 5 indicate, global warming may have a serious effect on the environment. The data that we need to resolve our uncertainty will almost certainly come from satellite measurements of surface melting on the ice sheet surface. You can find data of this type on websites such as NASA's.

Activity Box 5: Investigating sea-level change

For a densely populated area (e.g. southern England, northern Europe, the east coast of North America), find a map that shows the present-day coastline and the 50m contour. Photocopy or trace the map, and on your copy draw the 50m contour line as if it were the coastline, and shade in all areas below the 50m level as if they were ocean.

Your map now shows what the area would look like if sea level were 50m higher than it is today – i.e. what it would look like if most of the existing glaciers melted.

■ What is the greatest distance that the coastline would move inland in your area?

■ What major towns or cities would be lost to the sea in your area?

■ What towns or cities would find themselves on the coast in this new world?

Try making maps like this for different part of the world, and compare the potential effects of sea-level rise on different locations.

Information Box 3: Disappearing tropical glaciers

On mountain glaciers a small temperature increase can lead to a rapid rise of equilibrium line altitude and a rapid decrease in the size of the accumulation area relative to the ablation area. Tropical mountain glaciers have therefore been thought of as very sensitive indicators of global warming:

■ Kilimanjaro lost more than 80% of its ice between 1912 and 2004 and at that rate of recession its glaciers would vanish entirely by 2020.

■ The largest glacier on Mount Kenya lost about 90% of its ice during the last century.

However, the situation is far from simple, and some measurements have indicated that the retreat of tropical mountain glaciers such as those on Kilimanjaro correlate more with atmospheric moisture content than with temperature. Some observers have suggested that the decay of Kilimanjaro's ice is due not to global warming but to recent deforestation in the area around the mountain, which has reduced the amount of moisture being supplied to the atmosphere. Reduced atmospheric moisture means more sunshine and less snowfall.

Another suggestion is that global warming will increase humidity and cause a higher proportion of ablation to occur by melting rather than evaporation. Melting is about eight times more energy efficient than evaporation, so ice loss could occur more quickly in a warm humid atmosphere. The link between warming climate and shrinking glaciers is very complex.

You can find more information about Kilimanjaro's disappearing glaciers at the websites of University of Innsbruck Tropical Glaciology Group and the National Geographic.

Activity Box 6: Global warming, melting ice and sea-level change

Ice-shelf collapse might be an indicator of environmental change, but it does not contribute directly to changing sea levels. Ice shelves are made of floating ice, so just as melting an ice cube in a glass of water will not change the level of the water, so melting an ice shelf will not change the level of the sea. Try it!

■ Part fill a tub or glass with chilled water and float an ice cube in it. This is like a miniature model of an iceberg, and will behave in the same way as any floating ice mass, like an ice shelf.

■ Mark the water level on the side of the glass, and wait for the ice cube to melt.

■ Check the water level at intervals as the ice melts and after it has completely melted. You should find that the water level stays the same.

■ Now let the water warm to room temperature (e.g. 20°C) without letting any of it evaporate; you should see that the water level rises slightly inside the glass. Melting your iceberg did not raise the water level, but warming up the water from near zero to 20°C did.

Repeat this experiment, but instead of floating your iceberg in the water stand it in a separate container and then pour the meltwater into your glass at the end. What conclusions would you draw from this experiment in terms of the effects of melting of land-based and floating ice? (It makes a big difference whether you melt land-based or floating ice.)

Sea-level change is made even more complicated by the fact that the ocean basins and the continents themselves rise or fall in response to the amount of weight that they carry. An increase in water volume equivalent to 65m of sea level will actually not cause such a large rise, because the ocean basins themselves will sink downwards in response to the extra weight of water in them. Similarly, continents rise or fall by hundreds of metres in response to the growth and decay of ice sheets, so sea-level change at the coast is very difficult to predict (see Information Box 4).

Because isostasy occurs slowly but water volume changes quickly, rapid post-glacial sea-level rise caused by meltwater flowing to the oceans is generally followed in glaciated areas by a slow, long-term sea-level fall associated with isostatic recovery. The effect of glaciers on sea-level can be complex and it is important to remember that changes observed today might be caused either by changes in ocean volume related to recent changes in glacier extent, or by delayed isostatic responses to changes that occurred in glaciers long ago. Modern sea-level

Information Box 4: Sea-level change and isostasy

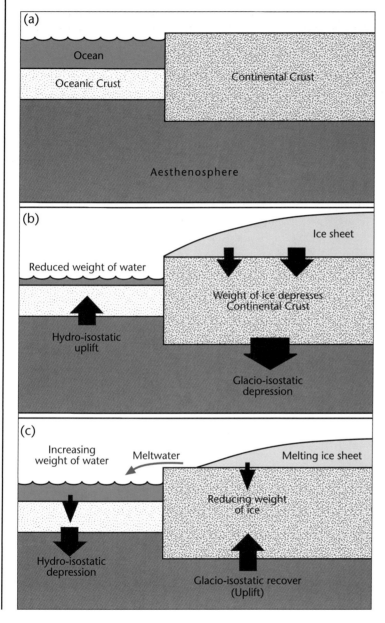

Figure 6: The effects of glacio-isostasy and hydro-isostasy.

Sea-level change is caused not only by changes in the volume of water in the ocean but also by *isostasy* (changes in the relative elevation of sections of the Earth's crust in response to the weight of ice or water on them). For example, if the Antarctic ice sheet were to melt, the water released would be equivalent to 66m of water distributed across the world's oceans, but the actual sea-level rise that would occur is modified by two phenomena:

1. Hydro-isostasy
The volume of water released into the ocean basins exerts a weight onto the crustal floor of the basin and depresses the basin floor into the asthenosphere below by an amount equivalent to about one-third of the depth of extra water, so a 66m rise following melting of the Antarctic ice sheet would be transformed into a sea-level rise of about 44m.

2. Glacio-isostasy
Glaciated continents are depressed by the weight of glaciers in the same way that ocean basins are by the weight of water. During glacial periods, ocean water volumes and global sea levels are low, but glaciated continents are depressed by the weight of glaciers so their relative sea level is higher than that experienced on non-glaciated continents. Upon deglaciation, removal of the weight of ice leads to isostatic rebound or recovery of the land.

change is not always caused by modern glacier change.

Glaciers, oceans and climate change
There is a complex, two-way relationship between glaciers and climate. Climate change has a major effect on glaciation because glaciers only form in certain climates. Low temperatures and high precipitation favour glaciation. Warmer or drier conditions cause negative mass balance and glacier decay. The situation is complicated when conditions get warmer but wetter, or colder but drier, and that's why the response of glaciers to climate change can be hard to predict. For example, a rise in global temperature might lead to increased melting at ice sheet margins but could also cause increased snowfall

in ice-sheet interiors. Global warming could thus cause ice sheets to thicken in the middle while they melt at the edges.

It is important also to realise that glaciers affect climate. Climate, ocean currents, atmospheric circulation, and the input of solar radiation to the Earth's surface are all part of the same system. For example, the growth and decay of large ice sheets disrupts the global energy budget by changing the albedo (reflectivity) of the Earth's surface. During glacial periods more of the surface both of the land and of the oceans is covered by ice; more of the sun's energy is reflected into space; and the sun's warming effect on the Earth is reduced. Glaciers also affect climate by releasing cold meltwater into the oceans. Variations in the amount of water that is released cause changes in ocean currents and consequently in global climate (see Information Box 5).

The history of the planet in ice cores

As glaciers are made up from layers of snowfall, they record the characteristics of the atmosphere from which the snow fell. The snow is made up of water molecules that reflect the chemistry of the world's water at the time it fell, and air trapped within the snow gets locked up as air bubbles in the glacier, preserving actual samples of the atmosphere from the

date the ice formed. If you go up to the edge of a glacier, chip off a bit of ice and watch it melt, you are releasing air that could have been locked away in those bubbles for hundreds, thousands, or even millions of years, depending on the age of the ice.

Because glaciers build up gradually over thousands of years, with the oldest ice at the bottom and the youngest at the top, scientists can drill down through the glacier to sample ice that formed at different points in history. Ice from deeper and deeper in the glacier is older and older and preserves a record of what the atmosphere was like longer and longer ago. In small glaciers the ice flows through so quickly that the oldest ice is only a few hundred years old, but in the large ice sheets of Greenland and Antarctica the ice near the bottom can be hundreds of thousands of years old. By analysing this ancient ice we can find out about climate change, about any prehistoric volcanic eruptions that affected the atmosphere, and even about industrial pollution at different times in history.

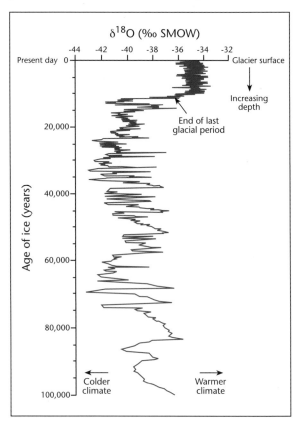

Figure 7: The oxygen isotope record from Greenland.

Information Box 5: Heinrich Layers

The Heinrich Layers are layers of debris that have been discovered in sediments on the north Atlantic seabed. The debris was deposited in layers by a series of short-lived but massive discharges of icebergs and meltwater that occurred at intervals of 5-10,000 years, between 14,000 and 70,000 years ago. These Heinrich events were caused by rapid advances and retreats of the Laurentide ice sheet in North America, and they caused sea-surface temperature and salinity in the Atlantic to drop sharply for periods of time. This completely changed the pattern of ocean currents, temporarily stopping the Gulf Stream that normally warms the North Atlantic, and increasing the temperature contrast between high and low latitudes. Immediately after each event, the circulation restarted and heat transfer from low latitudes resumed, causing abrupt warming in high latitudes. The impact of the Heinrich events on ocean temperature and salinity illustrates the potential impact of glacier fluctuations on climate.

Water comprises oxygen and hydrogen atoms bonded into H_2O molecules. Oxygen atoms can occur in several different varieties (called isotopes) with identical chemical properties but slightly different weights. The heaviest is Oxygen-18 (or ^{18}O) and the lightest Oxygen-16 (or ^{16}O). Water is made up of a mixture of the different isotopes, and the ratio of ^{18}O to ^{16}O atoms in a sample of water is determined by environmental processes, such as evaporation and condensation, that the water has undergone. It is possible to measure the isotopic composition of water, and thus work out its environmental history.

The isotopic composition of snow is controlled by factors such as temperature and ocean volume, so the layer-by-layer isotopic record of an ice sheet can provide a history of long-term climate change. Figure 7 indicates how the oxygen isotope ratio of snow falling in Greenland varied over the last 250,000 years. The information was collected by boring into the ice sheet and collecting a long core of ice for laboratory analysis. The youngest ice (most recent snowfall) is at the top, and the oldest at the bottom. The end of the last ice age about 10,000 years ago, and other key events in the Earth's history, show up clearly (see Case Study 2). Similar results can be obtained by measuring the isotopic composition of sediments built up slowly over time on the sea floor.

Case Study 2: Recognising volcanic eruptions and industrial pollution in ice cores

Anything that falls onto the accumulation area of an ice-sheet can be preserved in the ice. This includes volcanic ash and industrial pollutants. Figure 8 shows the volcanic dust content and lead content of ice from different depths in the Greenland ice sheet. Because different volcanic eruptions have different chemical characteristics, specific eruptions can be recognised from the chemistry of layers in the ice sheet. Where the date of the eruption is already known from historical records, the ash layer can be used to date the layer of ice.

The lead content of ice reflects human industrial activity. The amount of lead in glaciers increased enormously with the Industrial Revolution, but earlier industrial activity, such as the smelting of lead ore by the ancient Romans, has also been identified in ice cores.

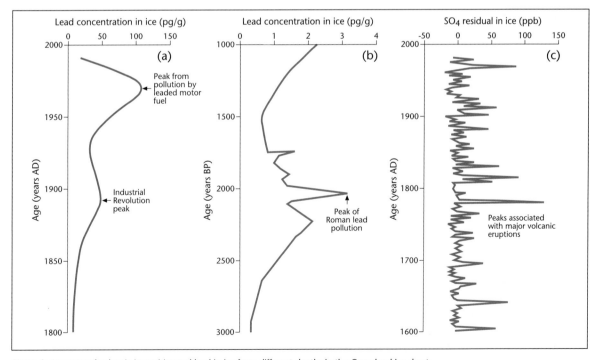

Figure 8: Amounts of volcanic impurities and lead in ice from different depths in the Greenland ice sheet.

Activity Box 7: Ice-core science

Put yourself in the position of a scientist studying an ice core extracted from the Greenland or Antarctic ice sheet.

■ What questions might you hope to answer by studying your core?

■ Apart from ice, what else might you expect to find in the core?

■ If the core came from the thickest part of your ice sheet, how long would you expect it to be?

■ How old would the ice at the bottom of the core be?

■ At what depth down the core would you expect to find the earliest evidence of industrial pollution?

■ How would you be able to work out the age of the ice at that depth?

To help you answer these questions look up the official websites of ice core projects such as EPICA (European Project for Ice Coring in Antarctica) and NGRIP (North Greenland Ice core Project – see pages 55-56).

Conclusion

Glaciers are an important part of a linked global system involving solar energy, climate, the hydrological cycle, sea level, and the circulation of the atmosphere and oceans. The system is complex, and is not yet fully understood, but glaciers can give us an insight into the dynamics, history and possible future of the physical environment. The role that glaciers play in that global system is determined by the way that glaciers behave, which is in turn controlled by their composition and structure. This composition and structure also reflects the way in which glaciers are created, and so can tell us a lot about the environment. Glacier structure is clearly an important issue, so we shall look at that next, in Chapter 3.

THE STRUCTURE AND ANATOMY OF GLACIERS

Glaciers come in different shapes and sizes, ranging from small mountain glaciers no bigger than a sports field (e.g. Palisade Glacier, California, USA – 1.6km²), through larger glaciers the size of a small country (e.g. Vatnajökull ice cap, Iceland – 8400km²) to ice sheets several kilometres thick that cover whole continents (e.g. the Antarctic Ice Sheet – 14 million km²). However, the physical properties of ice and snow are approximately the same for all glaciers, and all glaciers respond to the same controlling factors in the environment. Therefore, all glaciers have the same basic structure or anatomy.

Glacier shape and size

The size of a glacier is determined by how much ice it contains (ice volume) and how that ice spreads out over the landscape (glacier area). The volume of ice in a glacier is governed by the mass balance. The plan form of a glacier, or the way it appears on a map, depends primarily on how far the ice flows in any direction before it ablates. If the ice flows outwards equally in all directions from an accumulation area then the glacier will be circular in plan. However, many things can disrupt this flow pattern, and that's why glaciers can be many different shapes. For example, if more snow falls in one part of the accumulation area than others, or if ice flowing in one direction melts more rapidly, then the ice will travel further in some directions than others before it ablates. This is illustrated by the ice cap on the summit of the Ecuadorian volcano Cotopaxi, which extends further down the east flank of the mountain than the west because moist winds from the Amazon basin deposit more snowfall on the eastern side of the mountain (Figure 9). If ice flows faster in one direction than another, or if ice flow is funnelled in a particular direction by valleys in the local topography, this can also affect glacier shape. For example, ice building up in mountainous areas tends to be channelled into valleys. The shape of the glacier is constrained by the topography (see Activity Box 8).

Figure 9: The asymmetrical ice cap on the summit of the volcano Cotopaxi, in Ecuador. Photo: Peter G. Knight.

Activity Box 8: Glacier shapes

Find maps or air-photographs of the following glaciers and trace their outlines.

- Glacier 1: Hofsjökull, Iceland (ice cap)

- Glacier 2: Mer de Glace, France (valley glacier)

- Glacier 3: Brady Ice Field, Alaska (ice field)

What do you notice, and what can you say, about the glacier shapes in relation to the topography of the land?

Glacier stratigraphy

When glaciologists and geomorphologists try to understand how glaciers behave and how they affect the landscape, it is helpful to think about what goes on at the glacier surface (the *supraglacial* zone), what goes on at the bottom of the glacier where it rests on the ground (the *basal* zone and the glacier bed), and what goes on in the middle of the glacier inside the ice itself (the *englacial* zone) (Figure 10).

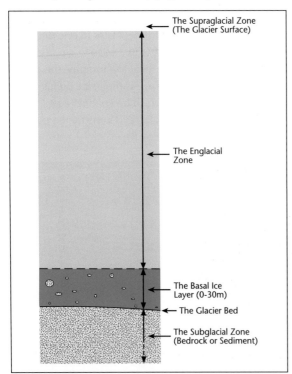

Figure 10: The main components of a typical glacier's stratigraphy.

The supraglacial zone

The glacier surface in the accumulation area is usually covered by snow all the year round. In the ablation area the surface may have snow on it during the winter, but for much of the year it will be bare ice, often with a lot of water and rock debris present.

Some of the debris on a glacier surface is transported upwards through the ice from the glacier bed, but most comes from supraglacial sources such as meteorites, volcanic ash, and rock falls from valley walls or from nunataks (mountains poking up through the ice). Many Icelandic glaciers, for example, contain prominent black layers of volcanic ash that melt out to cover the glacier surface with black debris. In the ablation zone, where ice melts, a lot of debris from within the glacier can be exposed at the surface and left behind to become a thick surface deposit. Many glacier margins are completely covered in rocky debris, and some even develop soils and have vegetation growing on them. Part of the Matanuska glacier in Alaska has so much debris on its surface that it supports a coniferous forest!

Surface water is important in the ablation area of most glaciers. Melting glacier surfaces commonly feature stream networks, ponds and lakes, and drainage connections to the interior of the glacier. Dendritic (branching) drainage patterns similar to those of land-based rivers can develop, and many river-like features can be identified in supraglacial streams. Where the ice in a glacier is very cold, surface-derived meltwater does not penetrate below the surface and the drainage network is confined to the supraglacial zone. On warmer glaciers the surface is often characterised by sinkholes in the ice known as *moulins*, which connect the surface to the internal drainage network.

Ice movement causes many distinctive features at the ice surface, in both the accumulation area and the ablation area. Crevasses are tensional features that form at the surface when the extending stress in the ice is greater than can be accommodated by ice flow, so the ice fractures. Crevasses are important because they indicate how the ice is moving: they are orientated at right angles to the direction in which the ice is extending most, and parallel to the direction in which the ice is compressing most.

Different levels of ablation of ice with different amounts of debris cover can lead to a variety of phenomena including perched boulders and dirt cones that form when debris protects ice beneath it from melting, and 'cryoconite' or melt holes that form when debris warms up in the sun and melts into the glacier surface. Whether debris on a glacier surface protects or melts into the ice depends on the thickness of the debris cover and its properties of heat absorption and transmission. Thin debris layers, or small individual particles, are more likely to transmit heat through to the ice than thick layers or large particles. Dark coloured clasts (pebbles and boulders) heat up more than light ones, and are more likely to transmit heat into the ice rather than reflect it away.

You can investigate these phenomena by carrying out the tasks described in Activity Box 9.

The englacial zone

The englacial zone is the portion of the glacier between the supraglacial zone (see above) and the basal zone (see below), and usually makes up the bulk of the glacier's thickness. Limited access to the englacial zone can be achieved by following crevasses or drainage channels from the surface, but most of what glaciologists know about the interior of glaciers comes from coring and from remote-sensing methods such as surface-penetrating radar.

The ice in the englacial zone has built up layer by layer from accumulation of material at the surface, and this is the ice that can be used for reconstructing

---

Activity Box 9: Perched blocks and cryoconite holes

Figure 11: Ice pedestal formed beneath a boulder on the surface of a glacier. Photo: Peter G. Knight.

- Prepare a big block of ice with a flat top in your freezer.
- On top of the block, place several pebbles ranging in size from small pieces of grit to large stones.
- Stand the ice, with the pebbles on top, in the sunshine or under a heat lamp.
- Repeat your experiment with different colours of pebble, light and dark.
- Try to find the critical size and colour at the boundary between pebbles that melt into the ice and pebbles that stand on pedestals.
- Record and illustrate your observations about the ways in which different sized and coloured pebbles behave or affect the ice.
- Describe the different kinds of ice forms that appear in relation to different sizes or types of pebbles (e.g. do any of them resemble the ice pedestal shown in Figure 11?)

Figure 12: An ice cliff at the margin of the Greenland ice sheet, showing a cross section through the ice in which foliations of bubble-poor (blue or grey) and bubble-rich (white) ice are clear. Photo: Peter G. Knight.

climate history (see Case Study 2, see page 19). There may be debris in the ice that accumulated at the surface, and the ice also contains air bubbles that were trapped as the ice formed. Flow of the ice often causes the air bubbles to be concentrated into bands up to a metre thick, separated by bands of less bubbly ice. This gives the interior of the glacier a striped appearance known as bubble-foliation (Figure 12).

The basal zone

The basal zone, often called the basal ice layer, is the part of the glacier in which the nature of the ice is directly affected by proximity to the glacier bed. Whereas the supraglacial and englacial zones comprise ice formed by compression and recrystallisation of snow that accumulated at the surface, basal ice can be formed either by the freezing of water at the bed of the glacier, or by metamorphism of surface-derived ice by thermal, strain and hydraulic conditions close to the bed. Because of this, the chemistry and the physical structure of the basal ice are different from the ice above. Englacial ice usually contains debris derived from the glacier surface, but the basal layer can contain large amounts of debris derived from the bed of the glacier. This affects the chemistry and structure of the ice, the way it moves, and its geomorphic potential. Where the glacier bed is melting, the basal layer may be thin or non-existent, but where water freezes to the bed, basal ice several tens of metres thick may develop.

The glacier bed

The bottom of a glacier is referred to as the *glacier bed*. The bottom face of the glacier is the glacier sole: it is like the bottom of a foot or shoe. The subglacial zone includes everything that lies underneath the glacier, including the substrate and anything at the ice–substrate interface. The character of the glacier bed is critical to glacier behaviour. Until quite recently, most glaciological models (see Chapter 6) assumed a rigid, impermeable, ice–rock contact. In fact, major parts of many glaciers rest not on bedrock but on sediment, often glacial till (a highly variable mixture of clay, silt, sand, gravel and boulders) produced by the glacier itself. This sediment, and therefore the glacier bed, is often permeable and deformable. Some glaciers have lakes underneath them. The largest subglacial lake yet discovered is Lake Vostok (see Case Study 3). While glacier beds can be accessed via boreholes, through cavities at the margin and via tunnels through the ice or subglacial rock, the glacier bed is difficult to observe directly, so much of our knowledge is based on observations of former glacier beds exposed by retreating ice (see Chapter 5).

Water in glaciers

All glaciers contain water. In cold glaciers water exists only in tiny amounts between ice crystals, but in temperate glaciers water forms huge lakes and river systems within the ice. The presence of water is critical to glacier behaviour and to the impact of glaciers on the landscape. Water is central to most mechanisms of glacier movement, provides a mechanism for transferring heat through the glacier, and plays a major part in glacier sediment production

Case Study 3: Subglacial Lake Vostok

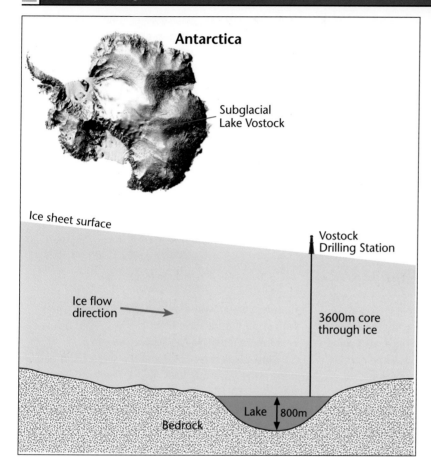

Subglacial Lake Vostok is an extreme example of water storage in the glacier system. It occupies an area of about 10,000km² (approximately the same area as Lake Ontario in Canada) and lies beneath ice 4km thick in East Antarctica.

Lake Vostok holds about 1800km³ of water and is fed by the melting of about 1mm/y of ice from the glacier bed. Thus, the residence time of water in the lake is estimated to be about 50,000 years. The mean age of the water in the lake, since it originally fell as snow at the surface of the ice sheet, is about 1 million years.

Figure 13: Position of Lake Vostok, Antarctica, and cross-section through the ice at the location of the lake.

and transport. The geomorphological and geological record of water activity associated with former glaciers has provided valuable clues as to their characteristics. Hydrological activity associated with glaciers is one of the major sources of glacier-related hazards. Volumes of water in glacier systems can be enormous. The glaciers of Alaska, in July and August alone, produce 2.5 million m³ of meltwater for each square kilometre of glacier surface. By contrast, some cold glaciers such as those in Antarctica accomplish most of their ablation by calving and sublimation, producing very little meltwater at all. (See Activity Box 10).

Glacier temperatures

Different glaciers have different temperatures, and the temperature of a glacier is one of the most important parameters controlling its character. Both the absolute temperature (whether the ice is at –5°C or –20°C, for example) and the temperature relative to the melting point (whether the ice is at the melting point or below it) are important. Ice temperature exerts a major influence on how ice flows, on the presence of water, on geomorphic processes and on the physical and chemical characteristics of the ice, as well as on the very existence of the glacier. The temperature at the bed of a glacier is an especially important control on glacier movement and the impact of glaciers on landscape.

Some glaciers are below the melting point throughout, and are referred to as cold glaciers. Others are at the melting point throughout, and are referred to as temperate or warm glaciers. Between these two extremes are many variations. For example, some glaciers are at the melting point except for a

---

Activity Box 10: Water inside glaciers

Try some experiments to see how water and ice interact. If a large group of you can work on this activity together, work in pairs and record your observations, then pool and compare your results. Be sure to wear gloves when handling cold ice; keep melting ice in a watertight basin or close to a sink where meltwater can drain away; and mop up any spilt water so that floors don't become slippery.

- Make two large blocks of ice, each with a cup-sized hollow on the top surface. Leave one block in the freezer and stand the other at room temperature for several hours until it starts to melt. Now get the 'cold' block out of the freezer and start the experiment.

- Pour water coloured with food dye into the hollows on the top of each block and observe how the water behaves. Usually, the water in the melting block (your model of a temperate glacier) will quickly find routes between crystals and through melt-channels into the ice. In the cold block (your cold glacier) the dyed water will probably take much longer to enter the ice, and will do so only when the ice begins to warm up.

- Experiment with different temperature ice blocks and different temperature coloured water. Try doing the same experiment with clay or sand in the water.

- What happens if you dip your 'cold' ice block into a bowl of cold water dyed with food colouring (to simulate a sub-glacial lake)? Repeat the experiment for a 'temperate' block.

- Record your observations, describing patterns of water and ice behaviour in relation to temperature.

surface layer that is subject to seasonal temperature fluctuations, while others are cold except for a layer of temperate ice near to the bed. Many glaciers exhibit spatial variations in temperature, with some parts being at the melting point and others being cold. Glaciers with several different thermal zones are sometimes referred to as polythermal glaciers.

We often assume that ice melts, and water freezes, at 0°C, but in fact the melting/freezing point can vary a little bit above or below this temperature. The melting point of ice varies with pressure and with the chemistry of the ice. Where ice is subjected to high pressure, as it is at the bed of a glacier, the melting point can drop a little way below zero, which means that ice colder than zero can still melt. Likewise,

turbulent water, water at high pressure, and water with impurities dissolved in it can be cooled below zero before it freezes (see Activity Box 11).

Most glaciers are coldest at the surface and get progressively warmer with depth. The temperature of the upper 10-15m of a glacier is affected by seasonal variations at the surface. The amount of seasonal variation decreases with depth, and at the bottom of this 10-15m layer the temperature is steady, and close to the mean annual surface temperature. Below that level, the effects of internal and basal heating lead to a general increase in temperature with depth. If geothermal energy were the only heat source, and the ice were stationary, temperature would increase steadily with depth at a rate of about 2.4°C per 100m

---

Activity Box 11: Testing the melting temperature of ice

- Make up several solutions of water and salt, or other solutes. Try using: distilled water; tap water; rain water; water with a little salt; water with a lot of salt; water with a little dilute sulphuric acid. Put a different mixture in each compartment of an ice-cube tray. Place the tray in the freezer, and check at intervals to see whether the different solutions freeze at different rates.

- Place two buckets of water to freeze inside a chest freezer with the lid open. In one bucket, safely set up an electric whisk or other device to keep the water turbulent. Compare the freezing rate of the turbulent water and the still water.

- Take two pressure sprayers of the type used for spraying fertilizers onto garden plants. Fill both with water, and pump one of them up to full pressure while leaving the other at normal atmospheric pressure. Place both in the freezer and see whether the high-pressure water stays liquid longer than the low-pressure water.

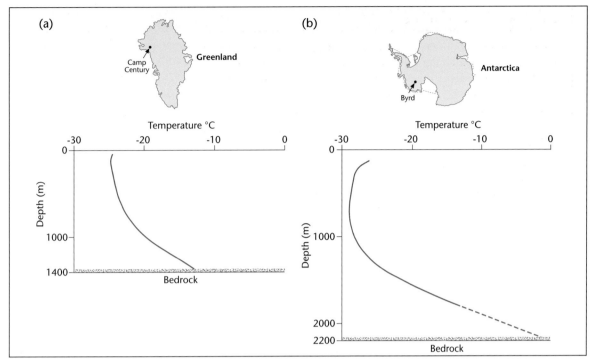

Figure 14: Temperature profiles through the Greenland ice sheet at Camp Century and the Antarctic ice sheet at Byrd Station.

depth until it reached melting temperature. Frictional heating from the movement if the ice, concentrated near to the glacier bed, increases the warming effect towards the base. Accumulation of cold snow at the ice surface and transport of surface ice downwards into the body of the glacier has a cooling effect.

Further complications are added by long-term climate changes and changes in glacier thickness or surface elevation, which alter the temperature of snow accumulating at the surface. For example, White Glacier, on Axel Heiberg Island in the Canadian Arctic, is marked by a temperature minimum 100-150m

Information Box 6: What controls the temperature of a glacier?

Temperatures in glaciers are controlled by heat sources at the surface, in the interior of the ice and at the base as well as by transfer of heat through the ice.

Location	Main controls on temperature
Surface	local climate (elevation, latitude, continentality)
Interior	internal deformation (friction) temperature of ice supplied from above and up-glacier heat conducted from warmer ice below
Base	geothermal energy friction between the ice and the bed

Melting of ice and freezing of water also affect temperature in a glacier. When ice melts it absorbs heat and cools down its surroundings. When water freezes it releases heat and warms up the area around it. Because the melting point of ice depends on pressure, and because the pressure can vary a lot from place to place beneath a glacier, a lot of melting and freezing goes on in most glaciers, so the pattern of heating and cooling can be complex. Ice melts at the melting temperature, which depends on pressure.

below the surface because of a period of colder climate prior to 1880. The warmest place in most glaciers is right at the bottom near to the bed. Many glaciers actually reach melting point at the bed so the bottom of the glacier melts to produce water (Figure 14 and Information Box 6).

Conclusion

All glaciers are a little bit different from one another in detail, but they share the same general characteristics. They can all be described in terms of their mass balance; they all have an accumulation area and an ablation area; they all have supraglacial, englacial and basal zones; and they are all affected by variations in temperature. Understanding the general characteristics that glaciers have in common allows us to appreciate the importance of the specific ways in which they differ. The following chapters look at the behaviour of glaciers to establish what the variations in detail between the appearances of different glaciers can tell us about exactly how those glaciers behave and how they are likely to respond to climate change.

CHAPTER 4

GLACIER MOTION AND FLUCTUATIONS

It is important to understand the difference between glacier motion and glacier-margin fluctuations. Glacier *motion* refers to the fact that the ice moves gradually through glaciers from the accumulation zone to the ablation zone. Just as water in a river moves downstream even though the river stays in the same place, ice in a glacier moves downglacier even if the position of the glacier on the map does not change. Glacier *fluctuation* refers to the fact that glaciers can expand and contract over time, changing the locations of their margins so that their size, shape and position on the map all change over time.

Glaciers in motion

The fact that glaciers move is really important. If glaciers didn't move, the world would be a very different place:

■ glaciers would be enormously thick but cover much less ground area

■ oceans would dry up as all the Earth's moisture would be trapped in polar ice sheets!

■ glaciers could not erode the land or transport sediment

■ there would be no glacial geomorphology.

Ice moves from the accumulation area to the ablation area, which means that in an ice-sheet glacier it moves outwards from the centre towards the edges, and in a mountain glacier it moves downwards from the top towards the bottom of the mountain.

Markers placed at different positions on a glacier surface and through the thickness of the ice from the surface to the bed, and resurveyed over a period of time, show the general pattern of ice flow. Markers at the surface move further than markers near the bed and, on a valley glacier, markers near to the centreline move further than those close to the valley walls. Markers placed on the surface in the accumulation zone are buried by accumulating snow, and follow a deep route through the interior of the ice; markers buried in the accumulation area emerge at the surface in the ablation area, with those from

higher up the glacier following the lowest route and emerging closest to the edge. Markers placed on the surface in the ablation area remain on the surface unless they fall into a crevasse or a moulin.

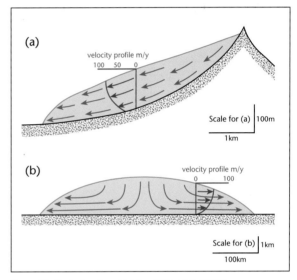

Figure 15: Typical routes of ice flow through a glacier.

Ice moves in response to stress (force) generated by its own weight. This is rather like runny dough spreading out on a tabletop as gravity tries to pull it downwards. In a glacier, the amount of stress affecting the ice is governed by the weight of the ice (which depends mainly on how thick the glacier is) and by the gradient of the ice surface. If you wonder why it is the slope of the *ice* surface, not the slope of the *ground* on which the ice rests, that controls the stress, think about water in a garden pond. If the water surface is horizontal, a slope in the floor of the pond won't make the water flow down it; but even if the floor of the pond is flat, any slope on the surface of the water will make the water flow until the surface is flat. You can't pile water up, and ice behaves in a similar way. If you make a big enough pile of ice (like an accumulating ice sheet) it will try to flatten itself out, and that is why glaciers flow!

Activity Box 12: Simulating glacier movement

You can reproduce many of the features of glacier flow with a model made of runny plaster of Paris (or blancmange mix, or a similar material). You can confine your 'glacier' in a piece of plastic guttering to simulate a valley glacier, or let it spread over a table to simulate an ice sheet. Try placing markers on (and in) a plaster 'valley glacier', tilting the gutter to make the glacier move, and observing whether the pattern of motion matches that described earlier in this book. For your plaster 'ice sheet' try placing large obstacles in the path of the advancing margin as you pour plaster onto the central 'accumulation zone', and watch how the obstacles (the 'topography') control the shape of the glacier margin. Compare the surface profile of ice sheets made with plaster of different consistencies. A more viscous plaster mix should make steeper, thicker glaciers, and a more fluid mix should make shallow-angled, thin glaciers. This simulates the effect of temperature on the viscosity of ice.

Be sure to wear protective clothing (gloves, overalls and goggles) when handling and mixing plaster of Paris. Work on sheets of plastic to avoid mess and dispose of materials safely after you finish.

Information Box 7: Stress and strain

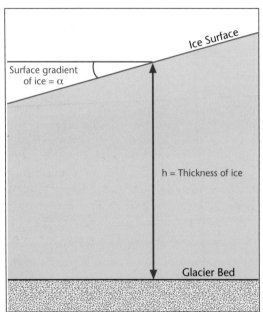

Figure 17: Cross-section through a glacier showing the terms used in the shear-stress equation.

The most important force causing motion in a glacier is *shear stress*. The term 'shear' here indicates that we are talking about a sideways force (making the ice move horizontally as opposed to a vertical force pushing the ice down onto the ground). The shear stress controlling motion of ice can be summarised by a simple equation:

$$T = \rho g h \sin \alpha$$

In this equation, T is the stress, ρ is the density of ice, g is the force of gravity, h is the thickness of ice above the point that you are calculating the stress for, and a is the gradient of the ice surface, as shown in Figure 16.

This equation shows that shear stress will increase if we have a thicker or steeper glacier. Because stress controls motion, we can now predict that steeper and thicker glaciers will move faster than thinner, less steep ones. Gradient (a) is very important. In the middle of the Greenland ice sheet, for example, even though the ice is 3km thick, it hardly moves at all, because the surface gradient of the ice sheet in the centre is very low.

The motion of the glacier, driven by stress, is called *strain*. The glaciologist John Glen discovered that in the deformation of ice there is a clear relationship between stress and strain, controlled by the hardness of the ice. Glen produced a famous equation sometimes called 'Glen's Flow Law':

$$E = A\tau^3$$

Here E is strain, τ is the stress, and A is a term that can be altered to reflect the hardness of the ice. This equation says that the strain will be equal to the cube of the stress multiplied by a number that depends on whether the ice is hard or soft. Glen also showed that the hardness of ice depends largely on its temperature. Ice close to the melting point is very soft (so we would use a high number for A in the equation), while very cold ice is very hard (so we would use a low number for A). From this we can predict that ice will flow faster in warm glaciers than in cold ones, and that a small change in the driving stress can have a big effect on the amount of motion. So, for example, slightly increasing the surface gradient of a glacier could make it go a lot faster.

Stress is not the only factor influencing glacier motion. The other is friction. If two glaciers experience the same shear stress, they might still slide at very different speeds if, for example, one rests on a smooth bed and the other on a rough bed, or if one is frozen to its bed while the other has a film of meltwater lubricating its base. Friction is a form of resistance, and glacier motion (strain) is controlled by the relationship between the stress that drives motion and the resistance that ... resists it!

Mechanisms of glacier motion

Glaciers move by three main mechanisms: internal deformation of the ice; sliding of the ice across its substrate; and deformation of the substrate beneath the ice. Until the 1970s, sliding and internal deformation were the only mechanisms that scientists recognised. The idea that deforming beds might contribute to glacier motion has been one of the most important new ideas in glaciology in recent decades. Measurements beneath a glacier in Iceland by G.S. Boulton and A.S. Jones in 1979 showed that 88% of the glacier's total motion was accomplished by deformation of the sediment on which the glacier rested. At the time this was a revolutionary finding, but substrate deformation is now a well established part of our understanding of how glaciers move.

Internal deformation (creep)

Because a glacier is a polycrystalline mass of ice (see Chapter 1, pages 8-9) it deforms under its own weight by a processes called 'creep'. This is a bit like modelling clay squashing when you squeeze it. In ice, the deformation involves: individual ice crystals re-aligning themselves within the ice mass; crystals melting and refreezing; and crystals actually deforming by motion of the atoms within the crystal. Together, these mechanisms allow ice, which at first sight seems like a solid material, to deform slowly in response to any applied force (see Information Box 7 for details of John Glen and his 'flow law' for ice). Because the stress on ice in a glacier is derived from the surface slope of the ice and the thickness (or weight) of the ice, it follows that ice deforms most quickly at the bottom of a thick glacier and where the ice surface has a steep gradient. Where ice is thin, or where the surface is flat, very little internal deformation will occur.

Basal sliding

The second important type of ice motion is sliding, which includes a range of processes by which ice moves across its bed. If both the sole of the glacier and the surface on which it rested were quite smooth, then sliding would be as simple as sliding this book across a tabletop. However, in any natural landscape the ground surface is rough at a variety of scales. There are tiny roughnesses in the rock surface; there a small bumps and hollows in the ground; and there are large hills and valleys. To 'slide' over this rough surface the glacier actually behaves in a quite complex way. It deforms around large obstacles by creep, and it overcomes small obstacles by melting a tiny amount of ice at a time, moving around the obstacle as water and refreezing on the downstream side of the bump. This process, known as *regelation*, is an important characteristic of glacier motion. The mechanism of glacier sliding over bedrock was explored by Barclay Kamb and Ed LaChapelle in a paper in the *Journal of Glaciology* in 1964 that is now widely regarded as a classic of glaciological research.

Substrate deformation

The third important mechanism of ice motion is substrate deformation. The substrate is whatever lies underneath the glacier. If the substrate is solid bedrock then it is unlikely to deform, but if the glacier rests on unconsolidated sediment (gravel, clay, or old glacial sediments, for example) then this sediment may deform under the weight of the ice. If the substrate moves, the glacier resting on top of it will also move. The same stress that drives creep and sliding, namely the weight of the ice and the gradient of its surface, drives motion in the substrate. The strength of the substrate depends on what it is made of, whether it is thawed or frozen solid, and whether there is a high water pressure within it. If the substrate is porous, unconsolidated and saturated, it will be weak, and subglacial deformation will be likely to occur. Glaciers on deforming substrates often move more quickly than those on bedrock and have lower (flatter) surface profiles.

Glacier velocity

Because there are so many different factors that can affect glacier velocity, glaciers move at a variety of speeds. Different glaciers move at different speeds, different sections of the same glacier can move at different speeds, and some glaciers can move at different speeds at different times. Glaciers are commonly classified into three types: fast, slow and surge (see Table 3). Ice in a fast-flowing glacier can move at rates of up to about 1m per hour, or several kilometres per year. However, this is quite unusual and most glaciers move much more slowly. Ice in a slow-flowing or 'normal' glacier typically moves only centimetres per day, or a few tens of metres per year. Some glaciers fluctuate periodically between slow and fast modes of flow. These are called *surge-type* glaciers. (See Case Study 4).

Table 3: How fast does ice move? Data from a variety of glacier types.

Glacier	Velocity (m/day)	Category
Jakobshavn Glacier, Greenland	19.0	fast
Columbia Glacier, Alaska	20.0	fast
Austdalsbreen, Norway	0.15	normal
Saskatchewan Glacier, Canada	0.35	normal
Lewis Glacier, Kenya	0.01	slow
Variegated Glacier, Alaska	0.50-65.0	surge type

Case Study 4: A surge-type glacier – Variegated Glacier, Alaska

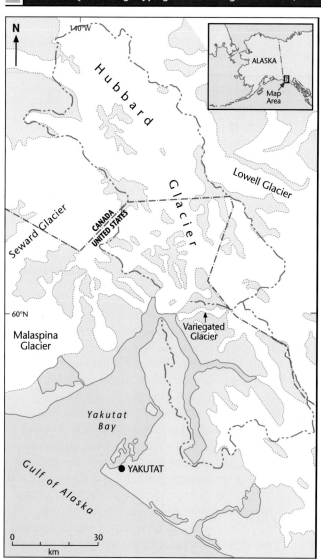

Figure 17: The location of Variegated Glacier, Alaska.

Every 20 years or so, Variegated Glacier in Alaska goes through a short surge period of very fast ice-motion (up to 65m per day) lasting less than a year. The surge is usually followed by a long quiescent period of about 20 years during which the ice flows at 'normal' speeds (a few cm/day).

After the 1964-65 surge, scientists studied the glacier intensively in the build-up to the next surge, and during the 1982-83 surge itself. They discovered that the surge occurred because of water building up at the base of the glacier. During the quiescent phase, meltwater escapes all year round through a network of tunnels at the glacier bed. However, the upper part of Variegated Glacier gradually thickens year on year, until it reaches such a thickness that the tunnel network can be squeezed shut under the weight of the ice. This occurs during the winter when there is not much meltwater flowing to keep the tunnels open. At the end of the winter in the year that the tunnels close, spring season meltwater cannot easily escape from the glacier, and so pockets of meltwater trapped at the bed build up to a high pressure. This water pressure reduces the friction between the ice and its bed, and causes the ice to start flowing quickly as it 'decouples' from the bed. This period of fast flow is the surge, and it continues until the meltwater manages to find a route out from beneath the glacier. During the surge, the rapid motion of ice through the glacier causes a thinning of the upper sections of the glacier. Because of this, the tunnels that are opened up by summer and autumn meltwater after the surge do not get squeezed closed the following winter, and the surge does not get repeated. The periodic repetition of surging is controlled by how long it takes the glacier to thicken up again to the critical tunnel-squeezing thickness.

Activity Box 13: Presenting glaciological information

Combine the information in Case Study 4 with other information about Variegated Glacier, or another surge-type glacier, from textbooks or on the internet in order to produce a poster, a web page, a PowerPoint presentation or a short talk about the glacier

You could include some of the following:

- A timeline showing the glacier's surge/quiescent phases through the last 100 years
- Diagrams illustrating the different stages of the surge cycle
- A graph showing the velocity of the ice at different stages of the surge
- Photographs of the glacier that you find on the internet, including images from satellite remote sensing
- Information about famous glaciologists such as Martin Sharp and Barclay Kamb who have worked on surging glaciers

Glacier fluctuations

Glaciers change in size at a variety of time scales, ranging from days to millions of years. As glaciers expand and contract, their margins advance and retreat. These fluctuations can influence climate and change the composition and circulation of the oceans and atmosphere. They can also affect human activity and alter the landscape. The landforms left behind by former glaciations provide evidence for the size and extent of ancient glaciers and therefore provide a way of reconstructing past climate changes.

Many glacier margins advance and retreat every year in response to seasonal changes in ablation. During the summer, ablation at the margin outweighs the forward movement of the ice and the margin retreats. Each winter, when ablation is low, the forward movement of the ice exceeds the rate of ablation and the margin advances. Variations in climate over periods of a few years or decades can also be reflected in equivalent periods of ablation-related advance or retreat. At longer time scales,

major fluctuations in glacier extent can be traced over centuries and millennia. At the longest time scale glaciers and ice sheets wax and wane, even to the point of disappearance and reformation, over periods of hundreds of thousands, or even millions, of years.

Fluctuations of glacier margins are caused by a mass imbalance at the margin. If the amount of ice supplied to the marginal area by flow from up-glacier is equal to the amount lost by ablation, then the position of the margin will remain stable. If supply exceeds loss the margin will advance. If loss exceeds supply the margin will retreat. Climate-driven variations in accumulation and ablation are, therefore, a major cause of glacier fluctuations. However, the response of the margin to changes in climate is not instantaneous. It takes time for the ice margin to retreat or advance to a new position. Changes in accumulation have a delayed effect on the position of the margin and for large ice sheets there might be a delay of thousands of years between an increase in accumulation at the summit and the corresponding

Information Box 8: The Little Ice Age

Between about 1500 AD and 1900 AD there was a period of globally cooler temperatures that is sometimes referred to as the *Little Ice Age*. During this period the climate was a few degrees colder than it is today. There are records, for example, of the River Thames in London freezing over regularly so that winter fairs could be held on the frozen surface. In the winter of 1780, New York Harbour froze, allowing people to walk from Manhattan to Staten Island. Famines were widespread and, for example, the population of European settlers in Greenland was wiped out entirely and thalf the population of Iceland died (see Activity Box 14).

Many of the world's glaciers responded to this cooler climate by expanding as ablation decreased. Since about 1900 AD, many glaciers have been retreating steadily from the maximum positions achieved during the Little Ice Age. That is why so many glaciers today are fringed by moraines (see Information Box 12) that mark the expanded limits from which they have recently retreated.

Activity Box 14: Documentary evidence for the Little Ice Age

Try to find evidence for the Little Ice Age in historical records or in art produced during the period. For example, in the writings of William Shakespeare, and in the art of painters such as Flemish painter Pieter Brueghel the Elder (Figure 18). For all the evidence that you can find, write down the nature of the evidence, identify the time periods that these records represent, and consider whether this 'evidence' of past climate is as reliable as, for example, modern meteorological recordings, or ice-core data, and why.

Figure 18: Representation of a frozen landscape from the Little Ice Age: *Winter Landscape with figures on a frozen river* by Brueghel, Pieter the Elder (*c.* 1515-69) (after) Bonhams, London, UK, Netherlandish, out of copyright.

reconstruct former climates, but changes in climate are not the only cause of fluctuations of glacier margins. Changes over time in the way a glacier moves (ice dynamics) can also affect ice supply and hence margin fluctuations. A periodically deforming subglacial sediment layer, or periodic floods of subglacial water, could lead to low basal friction, rapid flow, and advance of the glacier margin. These periodic advances (and subsequent retreats) are marked in the geological record by deposits of glacial sediment, but it would be a mistake to assume that the glacier fluctuations indicated temperature fluctuations. Glaciers are more complicated than that! Case Study 5 provides examples of how glacier fluctuations can be brought about by things other than simple climate change.

Conclusion

Glaciers move in response to stress driven primarily by

Case Study 5: Glacier fluctuations not directly caused by climate

a. Ice divide migrations: Myrdalsjökull

In ice sheets, ice flows outwards from the central accumulation area in response to the surface slope of the ice. Ice-sheet summits thus form ice *divides* that determine the direction of ice motion. However, the position of the ice divide can change through time. Directions of ice flow change: when this happens, different areas of the ice-sheet margin receive changed amounts of ice supply, and parts of the margin advance or retreat.

David Sugden and Andrew Dugmore of Edinburgh University studied the history of Myrdalsjökull ice cap in southern Iceland and showed that as the ice cap grew, local precipitation patterns changed and the accumulation centre shifted towards the southern flank of the ice cap. Their results showed that as the ice divide migrated, the size of the ice-catchment areas of outlet glaciers on the different flanks of the ice cap changed, and glaciers advanced or retreated accordingly. Glaciers to the south, including the glacier Solheimajökull (Figure 19), suffered shrinking catchment areas and retreated as the ice divide moved south, whereas outlets to the north experienced growth of their catchment areas and advanced. At other times, when the ice cap diminished in size, the reverse occurred, and Solheimajökull experienced an enlargement in its catchment area and advanced, while northern outlets experienced catchment shrinkage and retreat. The advance or retreat of individual outlet glaciers did not directly reflect climate change or the overall mass of the ice sheet, but depended on the changing position of the ice divide.

Figure 19: The glacier Solheimajökull, an outlet valley glacier that descends from the Myrdalsjökull ice cap in southern Iceland. Photo: Peter G. Knight.

b. Floating glaciers: Glacier Bay, Alaska

Floating glacier margins in fjords or bays are unlikely to be reliable indicators of climate and mass balance, because their advance and retreat is strongly controlled by topography. Glacier Bay National Park, in southern Alaska, is an area in which many glaciers terminate in the sea, and different glaciers within the Park are at present advancing and retreating at different rates. The ablation rate of a floating glacier is controlled by the amount of calving (the breaking off of icebergs) that takes place from its terminus, and this is determined partly by water depth and the width of the calving front. For an advancing glacier, the margin will advance until it reaches a position where the fjord is wide or deep enough to calve away all its supply of ice, even if this means advancing all the way to the open sea where the ice front can expand unrestricted. In the reverse situation, a retreating glacier will retreat along the fjord until it reaches a section where the amount of calving is reduced to match the ice supply. If the fjord does not get narrower or shallower, the margin will keep retreating. This means that a small climate change could lead to a huge advance or retreat, and for floating glacier margins the amount of ice advance or retreat is not related directly to the size of the mass-balance (or climate) change causing the fluctuation. This applies not only to fjord glaciers, such as those in Glacier Bay, but also to ice shelves, such as those in Antarctica.

(b)

Figure 20: The floating terminus of Jakobshavn Glacier in west Greenland. This is one of the fastest flowing glaciers in the world, and calves into its fjord to produce large numbers of icebergs. Photo: Peter G. Knight.

Activity Box 15: Glaciers and global warming

There is a great deal of debate about whether modern glacier fluctuations are caused by global warming. Many websites tend to assume that any kind of glacier retreat must be due to global warming, although we saw in Case Study 5 that in fact glacier retreat (and advance) can be caused by non-climatic controls such as topography and ice-divide migration.

- Visit websites about global warming and evaluate the way that they present glaciers as evidence for climate change. A good site to start with is the World View of Global Warming.
- Could the fluctuations of glaciers that are used as evidence of global warming be influenced by non-climatic factors such as topography?
- Do the websites also point out that some glaciers are advancing?
- Discuss with other students who have looked at the websites whether the evidence presented seems fair or whether it is biased.
- What additional evidence would help you decide whether glaciers are retreating because of global warming?

gravity and the surface gradient of the ice. Motion occurs by three main mechanisms: sliding, internal deformation and substrate deformation. Glacier fluctuations occur primarily in response to changes in mass balance. The principal driving forces are thus usually climatic, but ice-dynamic and geographical controls strongly influence the timing and the amount of fluctuation that results from environmental changes. Changes in positions of ice margins reflect the composite effect of several controlling factors, and cannot always be interpreted as simple indicators of climate change. Having considered how glaciers behave, and what factors control that behaviour, we are now ready to consider the effects of glacier behaviour on landscapes, which is the focus of Chapter 5.

GLACIERS ALTERING LANDSCAPES

Geomorphology is the study of landforms, landscapes and the processes that create them. Areas that have been glaciated usually have distinctive landscapes because glacial processes affect the landscape in distinctive ways. For example, glaciers erode rock into U-shaped valleys and deposit sediment to make moraines. Glaciers also affect landscape indirectly in more subtle ways. For example: drainage patterns can be disturbed when parts of a drainage basin are blocked by glaciers; coastal landforms and inland river systems far from the glaciated areas can be changed by glacier-induced sea-level change; and glacial meltwater can create fluvioglacial landforms that, although not directly created by ice, are nevertheless an important part of the glacial landsystem (*fluvioglacial* refers to water associated with glaciers). To understand glacial landforms we need to know about the processes that create them. Studying landforms can also help us to understand glacial processes. In this chapter we look at the key processes and some of their effects on landscape.

Glacial geomorphic processes

Glacial geomorphology involves erosion, entrainment, transportation and deposition. Glaciers erode the ground beneath them; they pick up or *entrain* debris; they transport that debris from place to place; and they deposit it as sediment. Landforms are created both where material is removed and where it is deposited, and the characteristics of landforms reflect the characteristics of the glaciers that created them. For example, fast-moving, warm-based glaciers generally produce different landforms from slow, cold-based glaciers. Geomorphologists use the link between glacier characteristics and landform types to work out what ancient glaciers must have been like by looking at the landforms they left behind. This is called 'inversion modelling', and is an important area of overlap and collaboration between glaciologists and geomorphologists.

Glacial erosion

Erosion is the removal of material by some moving agent such as wind, water or ice. In glacial geomorphology the principal agents are ice (glacial erosion) and water (fluvioglacial erosion). Erosion rates beneath glaciers vary from less than 0.01 to more than 100mm y^{-1}, and erosion can involve several different processes (see Information Box 9).

Glacial debris entrainment

Entrainment is the incorporation of debris into or onto the glacier, from either supraglacial or subglacial sources.

Supraglacial sources include:
- material falling or being washed or blown onto the glacier from surrounding land
- atmospheric fallout such as volcanic ash, or particles carried down onto the ice surface in snow or rain, including industrial pollutants.

Subglacial sources include:
- material eroded from the glacier bed or valley walls
- material frozen to the base from subglacial streams
- material frozen onto the base of floating ice from the underlying water.

Material entrained at the bed can be raised into the ice to form a basal layer (see Chapter 3). The main entrainment mechanisms, involving both attachment of material to the sole and the vertical transport of material from the sole into the basal ice, are listed in Information Box 10.

Glacial sediment transport

Material can be transported horizontally and vertically through glaciers by:
- The movement of the ice itself carrying the material with it
- Water transporting sediment through the glacier drainage system
- Glacial deformation of subglacial and proglacial sediments.

Information Box 9: Processes of glacial erosion

Abrasion is the scratching and polishing of bedrock by rock particles dragged across it by the glacier.

- Abrasion by individual clasts produces grooves or striations.
- Abrasion by 'rock-flour' polishes the bedrock surface.
- Abrasion rates are higher under faster-moving ice.

Fracture and traction describes what happens when moving ice becomes attached to the glacier bed and pulls away broken fragments. Traditional theory imagined a bond of frozen water sticking the rock to the ice and the ice 'plucking' the rock from the bed. However, ice doesn't stick to rock well enough to pull it apart, so plucking doesn't really work on solid bedrock and the concept of plucking is falling out of use. Modern theory argues that the crushing effect of ice passing over the rock, and variations in water pressure under the glacier, are essential to fracture bedrock and make erosion by traction possible.

Meltwater erosion can be mechanical or chemical. Mechanical meltwater erosion is similar to fluvial erosion in other environments except that water pressure in a glacier may be higher than atmospheric pressure. In geomorphology, chemical processes usually come under the heading of weathering, but dissolution of mineral material into water beneath a glacier can also be considered as an erosion process, because material is removed from the site of the reaction by the agent effecting the reaction (the water). Glacial environments have chemical erosion rates much greater than the global average because:

- solubility of carbon dioxide in water increases at lower temperatures, so glacial meltwater can become more acidic than warmer waters
- subglacial environments are usually rich in freshly-ground rock fragments that are chemically reactive and have a high ratio of surface area to volume
- water often passes very quickly through parts of the glacier system, which means that the water does not become chemically saturated but remains chemically reactive.

Activity Box 16: Abrasion and basal ice experiment

- Make a selection of ice blocks containing different types and amounts of debris, and see how effective each block is as an abrasive tool. Make the first block by freezing about 10cm of water in a plastic beaker. For the second block drop a handful of sharp gravel into the water before you freeze it. For a third block use clay or silt instead of gravel.
- Remove the ice from the beaker, and rub the bottom surface hard against a wooden surface as if you were sandpapering it.
- Describe the extent of abrasion up until the point when the ice begins to melt, and compare how much abrasion you get with the clean ice to the amount you get with different varieties of debris-rich ice.
- Be careful to wear protective gloves when handling cold ice, confine your abrasive activities to appropriate surfaces, and avoid making surfaces slippery with spilled water.

Activity Box 17: Freezing material to the base of a glacier

- Prepare a tray of wet sand and gravel completely saturated with very cold water to represent the bed of your glacier.
- Prepare two ice cubes: one kept in the freezer as cold as possible and one allowed to warm up to melting point. These represent your glacier.
- Push the two ice blocks into the wet gravel for a moment and then remove them again.
- Record your observations about entraining of the sediment from the bed onto the cold and the warm ice, and the level of friction apparent between each ice block and the glacier bed.
- Remember to use gloves when handling very cold ice!

Information Box 10: Basal entrainment mechanisms

Congelation
Congelation (the 'freezing on' of basal water to the glacier sole) commonly occurs where water flows into a cold zone at the bed or is 'supercooled' by sudden reduction of pressure.

Regelation
Localised pressure melting and refreezing of basal ice around bedrock bumps is an important mechanism for entrainment of debris. Because the temperature at which water freezes (or ice melts) varies with pressure, meltwater is generated on the upstream side of bed obstacles where pressure is higher and refreezes on the downstream side where pressure is lower. As the water refreezes to the base of the glacier, debris in the water, or eroded from the rock by fracture and traction, can be entrained into the ice.

Water-flow through the vein system
Water formed by pressure melting at the bed can be squeezed into the ice through the vein network between ice crystals, carrying with it fine silt or clay from the bed.

Shearing and folding
Layers of ice close to the base of a glacier can be folded or sheared by flow, and sediment from the bed can be carried upwards into the glacier along shear planes or within folded layers.

Apron over-riding
Ice or debris already existing beneath or in front of the glacier can be attached to the glacier sole to form part of the basal layer if the glacier advances over it (Figure 21). Proglacial ice sources include ground ice, lake ice, and glacier-margin accumulations of snow and ice known as 'aprons'.

Figure 21: Like many glaciers, the Greenland Ice Sheet contains large amounts of debris in its basal ice layer. Photo: Peter G. Knight.

In ice sheets most debris is transported in the basal layer because very little is supplied via supraglacial sources. By contrast, in glaciers surrounded by mountains much more englacial and supraglacial transport occurs. In supraglacial transport (and in englacial transport if particles are not touching each other) little alteration of the debris occurs. If clasts (pebbles and boulders) are transported in contact with each other or with the bed, processes of *comminution* (grinding down) tend to create characteristic particle shapes and particle size distributions that depend partly on distance of travel and style of transport. For example, *striations* (grooves and scratches) on clasts usually indicate transport at a sliding bed. In glacial sediments, characteristics of clast size, shape and orientation can be used to reconstruct transport pathways. The distribution of clasts of different lithology (rock type) in tills can also be used to reconstruct how rock has been carried down-glacier in ice flow.

Glacial deposition

Deposition occurs when material is released from the ice at the margin or the base of a glacier. Deposition may occur directly onto the ground, but many glaciers release sediment into water. The characteristics of glacial sediments can reflect both the processes of their deposition and also the processes of glacial erosion and entrainment by which the material was produced. The main mechanisms of deposition include:

- release of debris by melting or sublimation of the surrounding ice
- lodgement of debris by friction against the bed
- deposition of material from meltwater (fluvioglacial deposition)
- disturbance or flow of previously deposited sediments
- chemical precipitation.

Glacial process environments

Glacial landforms are often discussed in terms of the processes that create them: landforms of erosion or of deposition (e.g. in Addison, 1997). We can also consider landforms in terms of the locations within a glacier where they are formed: subglacial landforms or ice-marginal landforms, for example. Different parts of a glacier are associated with specific sets of geomorphic processes creating specific assemblages of landforms and sediments. This linking of location and process defines what we call a *process-environment*, and a process-environment together with its associated *sediment-landform assemblage* is sometimes referred to as a *landsystem*. Features associated with process-environments in different parts of the glacier system are discussed below.

Subglacial geomorphology

Subglacial geomorphology includes eroded bedrock, eroded and deformed sediment, deposited till, and fluvioglacial features. Erosional landforms include 'negative' (recessed) features such striations, grooves and troughs, and 'positive' (protuberant) features such as *roches moutonées* (asymmetric rock bumps). Landscapes incorporating such features include regions of *areal scouring*, where the ground is affected by widespread erosion, and regions of *selective linear erosion*, where local variations in subglacial conditions lead to abrupt local differences between areas of erosion and areas of no erosion. Depositional and deformational phenomena occur at all scales from vast till sheets to individual clasts. Many subglacial landforms are controlled by the movement of sediment and solutes in meltwater at the bed. Characteristics such as the thermal regime of the subglacial environment control the processes that operate, and hence the landforms and landscapes that are produced. Cold-based areas are likely to experience little sliding and minimal erosion. Warm-based areas are more likely to be associated with sliding and significant abrasion. Melting of debris-bearing basal ice is likely to cause deposition. The origin of some subglacial landforms, such as drumlins, remains unclear (see Information Box 11).

Geomorphology at the glacier margin

The main processes at the glacier margin involve the release of debris and water, and the reworking of previously deposited debris. Sediment can be released directly from the ice at the margin, transferred to the margin by subglacial deformation, or brought to the margin by water flowing from the interior or surface of the glacier.

The landforms created depend on whether the margin occurs on land or in water. On land, principal features include *moraine ridges* (see Information Box 12) and *outwash fans*. Material carried out of the glacier by water can form a low-gradient debris fan in front of the glacier. These *ice-contact fans* can thicken over time in front of a stationary ice margin so that when the ice eventually retreats a steep *ice-contact slope* is left to mark the former position of the margin, with the surface of the fan standing higher than the exposed ground up-glacier of the former margin. Where glaciers terminate in water, features can include subaqueous moraines and debris fans, as well as *fan deltas* that form when debris fans grow and emerge at the water surface. There is also a *distal proglacial zone* (an area some distance from the glacier that is still affected by it) in which sediment settles out from suspension in the water and rains out from icebergs drifting away from the ice margin (Information Box 12).

Information Box 11: Drumlins

A drumlin is a streamlined hill of subglacial sediment. Drumlins vary significantly in size, shape and composition, and their origins remain controversial. Several different origins have been proposed and different drumlins probably form in different ways. Some researchers suggest that drumlins are remnants of subglacial erosion by huge 'megafloods'. Other researchers suggest that drumlins are deposited by subglacial lodgement or meltout. Another widely supported model is that they are created by differential movement of areas of material within a deforming subglacial sediment layer. The drumlins may reflect areas of less mobile (possibly frozen) sediment while the intervening areas reflect areas where more rapidly moving material has been evacuated. Although drumlins are perhaps the most intensively studied glacial landform, our understanding of them remains incomplete.

Information Box 12: Moraines

A moraine is a landform created by the deposition or deformation of sediment by glacier ice. Different types of moraine reflect the different processes by which glaciers deposit and deform sediment in different parts of the glacier system. Moraines are composed of till, which is a highly variable mixture of clay, silt, sand gravel and boulders. Ice-marginal moraine ridges form by:

- sediment deformation in front of an advancing or oscillating ice margin (glaciotectonic or push moraines)

- dumping of sediment from the ice front over a protracted period at a stationary margin (dump moraines)

- squeezing of sediment from the subglacial deforming layer (extrusion moraines).

Figure 22: A moraine ridge at the edge of the Greenland ice sheet. Notice that there is a modern moraine being created by deposition directly from the glacier and an older, vegetated moraine ridge that was created several thousand years ago by an earlier glacial advance. Photo: Peter G. Knight.

Moraine ridges (Figure 22) provide a long-term record of the position of the margin. Glaciers that retreat episodically with periods of still-stand or minor re-advance can leave a series of moraines marking recessional positions of the margin. Glaciers in continuous retreat, without extended still-stands or re-advances, tend to deposit only unconsolidated sheets of debris with no ridges.

The geography, morphology and sedimentology of moraines can be used to reconstruct the geography and characteristics of former glaciers. For example, sediment characteristics reflect the source location of the debris: supraglacial debris is characteristically angular, while basally derived debris is typically faceted, sub-rounded and striated.

One of the most important aspects of the glacier margin is the way in which it controls the release of water and sediment to the proglacial zone. When glaciers lie behind fringing moraines much of the sediment produced at the margin is trapped in the moraine belt, but when glaciers have no such fringing moraines, sediment is more likely to be transferred into the proglacial system. Huge pulses of sediment might be input to the proglacial system as advancing glaciers break through fringing moraines.

Proglacial geomorphology and meltwater landscapes

The proglacial environment is outside the ice margin but still directly affected by the glacier, and especially by meltwater. Because they are fed by glaciers, meltwater streams are characterised by high sediment loads and highly variable rates of discharge. Fluvial (river) geomorphology is highly dependent on the discharge variability and the sediment load of rivers, so glacial

Figure 23: A glacier-fed meltwater stream in front of Solheimajökull in Iceland. Note the braided channel pattern, the broad flat-bottomed valley and the coarse bouldery sediment. Photo: Peter G. Knight.

Case Study 8: Glacial Lake Missoula

Glacial Lake Missoula was dammed up against the retreating edge of the Laurentide ice sheet in North America. At its maximum, about 15,000 years ago, the lake held about 2500km³ of water (more than Lake Ontario holds today) (Figure 24).

On several occasions this water was suddenly released from the lake and flooded across the landscape of what is now the north-western USA. The discharge of these floods was much greater than any modern flood: estimates based on geological evidence suggest that the flood might have flowed at as much as 20km³ of water per second, which is 20 times greater than the combined flow of all Earth's rivers today. The effect of these short-lived but massive floods was to create the distinctive eroded landscape that is now known as the 'channelled scablands' area of the north-western USA (Figure 25) and to dump substantial volumes of sediment into the Pacific Ocean.

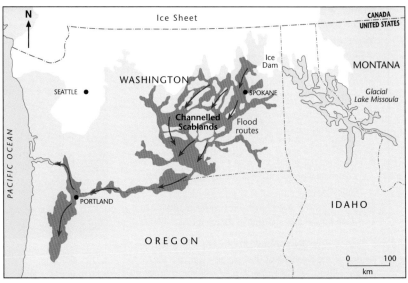

Figure 24: The position of Lake Missoula at the edge of the Laurentide ice sheet, flood routes and the channelled scablands.

Figure 25: Part of the 'channelled scablands' area that was eroded by the Missoula floods. Photo: Peter G. Knight.

streams create very distinctive fluvial landscapes. Glacier-fed streams are often braided and flow through broad, flat-bottomed channels lined with coarse, bouldery sediment (Figure 23).

Many glacier margins have lakes between the ice and the surrounding landscape. Water can be released suddenly from these lakes, for example when the glacier retreats, and any subsequent floods can have a severe impact on the landscape. After the glaciers are gone, the landscape can retain evidence of ancient lake beds, shorelines, and flood-eroded channels.

Activity Box 18: Glacier process-environments

For each of the glacier process-environments described above (subglacial, ice-marginal, proglacial and paraglacial), along with any other important process-environments that you identify from textbooks or with the help of a teacher:

■ Find images on the internet or elsewhere that illustrate the environment and the characteristic landforms and other features that occur within it. To help you get started you can find suitable images on my website (see Peter Knight's glacier pages).

■ Annotate your images with labels and information about the key features that you can identify. For example, on photographs of ice-marginal landscapes you could label the glacier margin, meltwater streams, moraine ridges, ice-dammed lakes, outwash fans, and so on.

■ Compile an 'atlas' of process-environments and their characteristic features. You could do this in the form of a *PowerPoint* presentation, with one slide for each environment, or a scrapbook or poster. If you are working in a group, each person could take responsibility for researching one process-environment, before combining your efforts to produce the atlas.

Paraglacial landscapes

Areas changing from glacial to non-glacial conditions, where surface features have not yet adjusted to the new postglacial environment, are referred to as *paraglacial*. For example, subglacial meltout till that was stable when deposited beneath the debris-rich basal layer of a warm-based glacier may be rapidly eroded when exposed to surface processes as the glacier retreats. Deglaciation can cause instability and rapid erosion lasting until a new equilibrium is established between surface materials and the postglacial process-environment. Paraglaciation is an important concept because many areas of the world are currently in a paraglacial condition following glacial retreat.

Glacial landscapes in the British Isles

Every part of the British Isles has been affected by glacial or proglacial environments during recent geological history. Areas such as the mountains of Scotland, the English Lake District and North Wales are famous for spectacular features of upland glaciation including corries (bowl-shaped hollows) and U-shaped valleys. Equally interesting are the more subtle features of lowland glaciation that

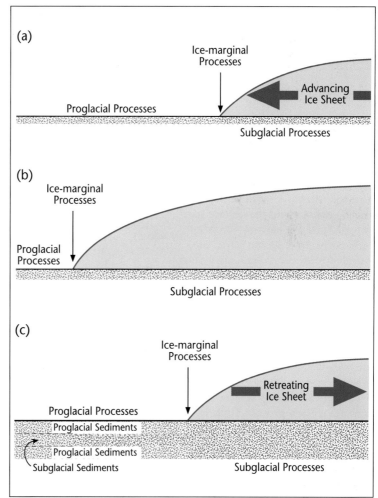

Figure 26: Glacial processes occur in sequence over time as the glacier changes position, leaving a record in the landscape of different processes acting at different times in the same location.

extend over much wider areas. These include the drumlin fields of Ireland, the till plains of central England, and widespread evidence of meltwater erosion and deposition. It is important to remember that the British ice sheet, like all ice sheets, grew, advanced, reached a maximum, and then retreated over a period of many thousands of years. Most areas therefore experienced a range of different glacial process environments over time: initially they were proglacial; then the ice approached and they became ice-marginal; the ice advanced further and they became subglacial; then the ice retreated again and the sequence was reversed. The landscape was thus affected over time by a sequence of different processes, with evidence of each stage being retained in the landscape (Figure 26).

Case Study 9: North Staffordshire

North Staffordshire is typical of many areas of Britain where glacial landforms are not spectacular but nevertheless play a major role in the landscape. Between about 20,000 and 15,000 years ago the southern margin of the British ice sheet progressively retreated north across the region. At one time, the margin ran through the area just west of Stoke-on-Trent as shown in Figure 27.

Ice flowed into the area from the north-west, and was blocked by the higher ground around the south-western edges of the Pennines. The configuration of ice and pre-existing topography created a distinctive landsystem of ice-marginal features including: an ice-marginal moraine ridge (the Woore moraine stretching westwards along the ice margin towards Wales); ice-dammed lakes such as Glacial Lake Madeley with sandy deltas where glacial streams flowed into them; and meltwater outwash valleys such as the Whitmore trough.

Today, the remnants of this landsystem can be seen, for example, in the flat land of the former lake bed at Madeley, the broad flat-bottomed valley at Whitmore and the hummocky ridge of the moraine at Woore. The whole landscape is now agricultural and well populated, and the glacial history is not obvious, but the underlying shape of the landscape is glacial in origin (Figure 28) (see also Case Study 11, page 5).

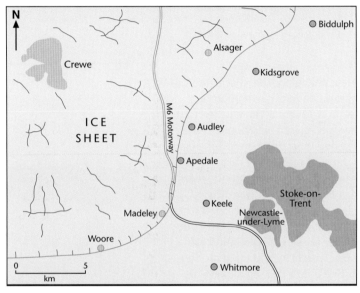

Figure 27: The position of the ice margin in what is now north-west Staffordshire about 18,000 years ago.

Figure 28: Staffordshire countryside today, with subtle legacies of the ancient ice sheet. Photo: Peter G. Knight.

Activity Box 19: Glaciation in your area

Wherever you live, glaciation has probably affected your local area. Using sources listed at the end of this book and/or other sources, try to find out all you can about the glacial past of your local area and the landforms that have been left behind.

- Find out if glaciers ever occupied the area where you live.
- If so, describe the type of glacier that existed, give dates when it existed and explain how it affected the landscape (e.g. Did it erode valleys? Did it deposit sediment? Did its meltwater erode valleys or make lakes that left a mark on the landscape? Are there any unexplained landforms?). Wherever possible include extracts from Ordnance Survey maps (at a suitable scale), images and graphics in your report to illustrate your text.

You could set up a web page to explain the formation of these features. Examples of student work can be found on school websites (e.g. Holgate School). However, not all school websites are this good, so be careful not to take everything on such websites as being reliable information: remember that they might have been created by students who are only just starting to learn about glaciers.

Conclusion

Glacial geomorphology arises from glacier processes, and can therefore be used as an indicator of such processes. Glacier processes such as sliding, meltwater production and the formation of basal ice by debris entrainment are directly related to geomorphic processes. Landforms therefore provide clues about the glaciers that created them. The study of landscapes is valuable in its own right, but it is also a valuable tool in glaciology. New understanding of processes such as substrate deformation has changed the way we think about glacial landforms, and there are still unanswered questions even about such common landforms as drumlins. The landforms that they leave behind are only one of the ways in which we can experience the effects of glaciers in our everyday lives. Glaciers can affect human activity in many ways, and we will consider some of those ways in Chapter 6.

GLACIERS AND HUMAN ACTIVITY

Glaciers affect people in many ways: some good, some bad. Although only a small proportion of the world's population actually lives close to a glacier, many more people live close to glacier-fed rivers, or live in houses built on glacial sediments. Everyone lives with an atmosphere, climate and oceans strongly influenced by glaciers.

Whether something is a hazard or a resource depends on the point of view of the people it affects. For example, meltwater is a problem for the bridge-builder but a boon for the hydropower engineer. The extent to which hazards can be mitigated or resources exploited depends partly on the level of technological development of the society concerned. For example, glacial resources such as ice for refrigeration, which are valuable to low-technology societies, are largely irrelevant in a high-technology society with electrical refrigeration. Economic development of increasingly remote areas increases human exposure to glacier hazards. For example, the way that ski tourism brings people into contact with glaciated environments provides evidence of some of the potential economic value of glaciated areas, but also provides potential victims for the natural hazards of those areas.

Glacier-related hazards

Ice avalanches

Where glaciers terminate on steep slopes, as is the case in many mountain areas, ice can break away or 'calve' from the snout (the end) and plunge downslope. Ice avalanches sometimes occur together with landslides or snow avalanches, but the largest known 'pure' ice avalanche, from Mount Iliamna, Alaska, in 1980, involved about 20 million m³ of ice. In 1597 an entire village near the Simplon Pass, Switzerland, was buried beneath an icefall, and in 1965 ten million m³ of ice fell from the Allalingletscher in Switzerland, killing 88 people working on the construction of a hydro-electric power plant in the valley below. Ice avalanches from Nevado Huascaran, the highest Peak in Peru's Cordillera Blanca, caused massive devastation twice in the twentieth century. In 1962, approximately 4000 people were killed by an

ice avalanche when 2 million m³ of ice fell 3000m from the peak into the populated valley below. The cascading ice reached speeds of more than 100km/h and buried nine villages in icy rubble up to 20m thick. Bodies were found as far as 160km downstream at the Pacific coast. In 1970, a combined rock and ice avalanche from the same mountain buried the town of Yungay, killing about 18,000 people.

Floods

Glaciers sometimes release large amounts of water in the form of outburst floods, also known as 'jökulhlaups' (their Icelandic name). These floods are among the most frequent and devastating of glacier hazards. They occur when water stored within, or dammed up by, the glacier is suddenly released, or when large amounts of ice are suddenly melted by volcanic eruptions. In Iceland, for example, more than 80 subglacial volcanic eruptions have been reported since the country was settled. In 1918 a jökulhlaup from Myrdalsjökull ice cap, Iceland, had a discharge equivalent to three times that of the Amazon. In southern Iceland on 5 November 1996, meltwater that had been produced as a result of a subglacial eruption beneath the ice cap Vatnajökull emerged as a jökulhlaup from the snout of the outlet glacier Skeiðarárjökull. Discharges of around 45,000m³/s swept away the roadway and bridges from a section of Iceland's main national ring-road, causing the Icelandic Prime Minister to say that 4 hours of flooding had knocked back the country's road-building programme by 20-30 years.

In most cases, the magnitude of major jökulhlaups is such that attempts at engineering defences are useless. Nevertheless, engineering measures can be taken to reduce the magnitude or likelihood of flooding. For example, the Gietroz Glacier in Switzerland advanced from a tributary valley to block the Val de Bagnes and create an ice-dammed lake in 1549, 1595 and 1818. In 1595 the outburst flood that was caused when the lake overflowed killed 500 people. When the lake formed again in 1818, engineers cut an artificial channel across the dam, and succeeded in draining about one third of the lake water before the lake burst. Nevertheless, 50 people perished in the resulting flood.

Mudflows and lahars

In many cases, water released from a glacier combines with loose sediment from moraines or other proglacial materials to produce mudflows. These are especially common in two types of situation:

- when moraine-dammed lakes burst and incorporate debris from the moraine into the flow; and
- when volcanically triggered jökulhlaups combine water with ash and unconsolidated volcanic materials to produce glacio-volcanic mudflows known as lahars.

One of the most deadly and widely known glacier-related disasters ever to have occurred was the lahar caused by melting of snow and ice during the eruption of the volcano Nevado del Ruiz in Colombia in 1985. About 10% of the volcano's ice cap melted, and the meltwater combined with volcanic ejecta to create lahars that flowed down several separate channels off the volcano. Peak flow in one channel reached 48,000m³/s, with a velocity of 38km/h and a wave front about 40m high. The lahar flooded through the town of Armero at about 30km/h, killing more than 20,000 people.

Case Study 10: Moraine-dammed lakes and mudflows in Peru

Recession of glaciers in the Cordillera Blanca of Peru after 1927 has resulted in the formation of substantial lakes between retreating glacier tongues and their fringing Little Ice Age moraines. As the lakes filled, the moraines became unstable and liable to collapse either because of the pressure of water or because the water level rose to a point where waves caused by ice-falls into the lake could overtop the dam. Some of these lakes drained catastrophically, incorporating material from the dam to course downhill as mudflows. In 1941 a flood killed 6000 people and devastated the town of Huaraz.

Immediately afterwards, work was started to mitigate the hazard of the Cordillera Blanca lakes. The floods have been successfully prevented by the installation of artificial drains and canals through the moraines. These artificially restrict lake depths and prevent dam failure. Today, the lakes and glaciers of the Cordillera Blanca are monitored by satellite (Figure 29), but differences of opinion between different experts in interpreting the threat suggested by satellite images has caused controversy.

A discussion of this problem features on the web pages of the Tropical Glaciology Group at Innsbruck University.

Figure 29: The area around Huaraz in Peru. Photo courtesy: NASA/JPL-Caltech.

Figure 30: A small iceberg off the west coast of Greenland. Photo: Peter G. Knight.

Icebergs

Icebergs pose a threat to shipping, oil platforms and sea-bed installations such as pipes and cables. In the northern hemisphere the most problematic icebergs are those that drift south into the North Atlantic shipping lanes. The main sources of icebergs affecting this area are glaciers on the west coast of Greenland (Figure 30). Up to 2500 icebergs per year drift southwards from Greenland towards the Grand Banks area east of Newfoundland. The area is sometimes referred to as 'iceberg alley' and in that area alone, between 1882 and 1890, 54 ocean liners were reported sunk or damaged in collisions with icebergs. The most famous victim of a North Atlantic iceberg was the SS *Titanic*, which was sunk with the loss of more than 1500 lives in 1912.

Sea-bed installations are subject to iceberg damage even at substantial distances from land, as even in water up to 200m deep, large icebergs can plough furrows in the sea bed and rip up cables and pipelines. Attempts to destroy icebergs by bombing and other means have largely been unsuccessful, but attempts to tow them for short distances to divert them from collision courses with fixed installations, such as oil rigs, have met with some success (and sometimes appear in the media, Activity Box 20). The first iceberg-proof fixed oil rig, the Hibernia, was established in the Grand Banks area off Newfoundland in 1997 at a cost of about US$5 billion. The rig is designed to withstand the impact of icebergs up to 6 million tonnes, but any icebergs approaching within 9.6km of the rig are towed aside to minimise the risk of collision.

Activity Box 20: Glacier-related hazards in the news

- For a period of one week, scan the international news every day to identify any items related to glacier hazards. (As well as obvious items such as floods and ice avalanches, watch out for indirect hazards such as sea-level rise.)
- Using these news items and other information in print (e.g. Richardson and Reynolds, 2000) and on the internet (e.g. Oxford Brookes University), produce either a glacier-hazards news magazine for the week in question, or a leaflet for people living in affected areas.
- Your magazine or leaflet should include advice about how people in affected areas should respond to the hazards.

Glacier-related resources

People have made use of glaciers and glacier-related phenomena since prehistoric times, and glaciers continue to be a resource in a variety of ways today. Some of the main resources associated with glaciers are discussed below.

Refrigeration

One of the earliest uses of glacier ice, and one that continues today in some areas, was as a refrigerant. Before refrigerators were invented, Norway exported ice to other European countries, and a substantial trade in North American lake-ice flourished in Europe in the second half of the nineteenth century. Icehouses built to supply the kitchens of the European aristocracy were stocked with ice from distant glaciers. In the 1850s, ice was transported from Alaska to California, and small icebergs from southern Chile were transported as far north as Peru for use as refrigerant.

Water supply

Glaciers are an abundant source of the most basic of human resources: water. Meltwater from glaciers is the source of many of the world's major rivers. Meltwater can be used for irrigation, for industry, to produce power, and for drinking and domestic purposes. Glaciers are especially valuable as a water source since they produce most water in hot, dry weather, at times when other sources, such as precipitation, are at a minimum. Some major cities, such as the Bolivian capital La Paz, derive their water supply almost entirely from glacier-related sources. In the USA, the Arapaho Glacier produces about 260 million gallons of drinking water per year for the city of Boulder, Colorado.

Some drinks companies have exploited the image of glaciers to promote sales of bottled water derived from glaciers. You can see examples of such products at commercial websites such as Alaska Glacier Refreshment.

Power

One of the major uses of water derived from glaciers is in power generation. In some countries, such as Greenland, the total human power requirement could theoretically be met by hydroelectric power fed by glacier meltwater. Norway derives all its electricity from hydro-electric power schemes, many of which are supplied by glacier water. The Massa hydro-electric power station near Brig in Switzerland runs largely off summer meltwater from the Grosser Aletschgletscher. Meltwater produced in summer is stored in reservoirs and released into the hydro-electric power station in winter. In some cases, water is fed into the power schemes directly from the glacier. For example, when the Mauranger power station in south-west Norway was designed, a sub-glacial water-intake was built at the bed of the glacier Bondhusbreen well above the level of the snout. A tunnel was cut through the bedrock beneath the glacier, and water intakes were drilled upwards to the glacier bed beneath ice 160m thick. Water is captured and fed to the power-station reservoir via a tunnel system. However, there are problems in the use of meltwater streams: the discharge is variable, difficult to predict, and tends to carry large amounts of sediment. Sediment in the water can block or erode water intake structures, can fill reservoirs quickly with sediment, and can cause rapid wear of pumps and turbines.

Icebergs

Icebergs, being composed of fresh water, offer a potential water resource if they can be transported to areas where water is needed. The cost of processing water from a captive iceberg would be less than the cost of recovering freshwater from ocean water by desalinisation. This is a good example of how the definition of a resource depends on human perception: if the need is great enough and a market exists, the cost and technical difficulty of exploitation can be faced. One major problem would be to transport the iceberg. Not only would the physical manipulation of an iceberg be an awesome task, but the iceberg would be subject to melting and fracturing en route. Problems could also arise from thermal pollution of waters en route, and at the destination, by the cold fresh water produced by the melting iceberg. Changes in local microclimate at the destination site could also occur. Problems would exist at the destination for docking and processing the iceberg. Large icebergs have a draught (depth underwater) of as much as 200m, and could not be towed close to shore. Around most of the south coast of Australia, for example, icebergs could be brought no closer than about 35km and would have to be moored and processed at sea. The use of icebergs in this way has not actually been put into practice yet, but is a real possibility if future climate change brings drought to more populated areas.

- Research the advertisements of holiday companies and national tourist boards to identify the different ways in which glaciers and glacial landscapes have been exploited by the tourist industry.
- Identify which aspects of glaciation provide the greatest attraction for tourists, and which aspects seem to be ignored by tourism companies.
- List as many different commercial companies as you can that base their business on glaciers. Consider tourism companies as well as equipment suppliers, caterers, etc.
- Compare ways in which advertisements use glaciers in different ways to appeal to different customer groups (for example younger and older age groups).

Waste disposal

It has been suggested that glaciers could be used as natural dustbins for radioactive industrial wastes. These wastes need to be kept out of contact with the biosphere for periods of up to 250,000 years. In the early 1970s, the International Atomic Energy Agency considered the possibility of burying waste beneath the Antarctic ice sheet, but the idea was rejected on the grounds that the necessary isolation from the biosphere could not be guaranteed. The potential consequences of accidental leakage from glacier storage are serious, because glaciers are directly linked to the rest of the global hydrological system (see Chapter 2). Further research into the issue has continued, however, as the problem of long-term storage of waste remains unsolved.

Materials and mineral tracers

Glacial activity produces materials such as sand and gravel. Many formerly glaciated areas such as Britain are rich in glacier-related sand and gravel deposits that are a major industrial resource. In Britain alone the building industry uses nearly 300 million tonnes of sand, gravel and aggregate each year. The distribution

of these materials reflects the geography of former glaciers and the geography of glacial geomorphic processes. For example, fluvioglacial landforms such as eskers, kames, and outwash fans, all of which are landforms deposited by meltwater from glaciers, are easily identifiable sand and gravel sources.

The distribution of glacial deposits can also be used to help locate mineral resources. In areas where bedrock is obscured by a surface covering of glacial sediment, mineral prospecting by direct investigation of bedrock is difficult. However, the location of specific lithologies or mineral deposits in the bedrock can be worked out from the distribution of material in the overlying deposits. The concentration of a particular material in till is usually highest within a few kilometres of the source, and then declines with distance in what is called a 'dispersal fan'. Locating the head of this dispersal fan by sampling the surface material can help in prospecting for new mines.

Glaciers and tourism

Since the mid-nineteenth century, glaciers have provided the basis for particular types of tourism, and now form the cornerstone of some local economies. For example, the tourist industries of Greenland and Iceland are in large measure based on their glacial landscapes. National Parks around the world are based either on present-day glaciers (e.g. Glacier National Park, Montana, Figure 31) or on landscapes left behind by ancient glaciers (e.g. the Lake District National Park in the UK) (Activity Box 21).

Ancient glaciers and the modern human landscape

One major effect of glaciation from a human point of view is the creation of new land surfaces by the deposition of glacial and fluvioglacial sediment. The characteristics of these deposits are important in engineering and construction because they provide the ground on which we build and many of the

Figure 31: The entrance to Glacier National Park, USA.
Photo: Peter G. Knight.

Case Study 11: The Apedale Valley

The Apedale Valley in North Staffordshire (see also Figure 27 and Case Study 9) is famous for its rich industrial heritage, with a history of mining and manufacturing stretching back to Roman times and incorporating some of the major innovations of the industrial revolution. This industrial heritage, however, is largely based on the area's geological and glacial heritage.

The geological underpinnings of the coal and iron industries are obvious, but the glacial history of the area has also been instrumental in establishing the modern industrial landscape. The overall form of Apedale was created by glacial meltwater. This spilled into the valley when the southern part of the British ice sheet occupied north-western England and terminated right on the watershed at the head of the valley. The meltwater was responsible for the broad, flat-bottomed trough that is still visible in places in spite of major industrial re-landscaping (Figure 32). The meltwater that carved out the valley also deposited sand that was washed out of the ice sheet.

Figure 32: Apedale today reflects a long and complex history stretching back through industrial activity, Roman settlement and ancient glaciation. Photo: Peter G. Knight.

Thousands of years later the sand was extracted for industrial use. Its extraction created a deep pool known as the Blue Lagoon, which was a prominent feature in the local topography for much of the twentieth century. The lagoon has now been filled in and returned to the sort of broad, flat landscape that would have existed immediately after the glacial meltwater stopped flowing. The glacial history and meltwater origin of the valley have influenced the valley's shape, the occurrence of sand and gravel in its floor, its industrial history and even its local ecology. The community country park that has been established in Apedale helps visitors to appreciate and understand all these aspects of its heritage. The visitors who come to Apedale today to explore and enjoy the natural and industrial landscapes are another example of the way that ancient, glacial processes can affect modern human activities.

materials that we use in building, such as sand and gravel.

Deposits related to specific glacial processes have specific properties. Grain size distribution, consolidation, jointing, and clast fabric are controlled by glacial processes, and influence properties such as bearing capacity, plasticity, settlement, slope stability, ease of excavation, and value as fill or construction material. For example, subglacial lodgement tills are likely to be difficult to excavate but to have good bearing and stability characteristics (good for building on). By contrast, ablation tills and fluvioglacial materials are likely to be easier to excavate but less stable (bad for building on). Waste disposal in formerly glaciated areas also requires knowledge of the properties of glacial sediments. For example, the inhomogeneous nature of glacial till, incorporating both impermeable clay-rich sediments and permeable lenses of coarser material in very small areas, make it unreliable as a barrier to groundwater flow, and thus unsuitable as a substrate for a waste tip without an artificial lining.

Activity Box 22: Advertising your local glacier features

- Identify a local tourist attraction such as a country park, National Trust site or other amenity near to where you live that has a connection with glaciation.
- Gather information about the site from brochures, local information centres or tourist websites.
- Visit the site and look for evidence of glaciation.
- If possible, arrange to interview staff and visitors at the site to see what they know about the glacial history of the area.
- Design a display board or leaflet that will educate visitors about the glacial history of the site. Alternatively, you might choose to devise a self-guided walk or a guide-book to the glacial features.

Even in areas such as Great Britain, where the last glaciers disappeared thousands of years ago, their effects can still be seen in the human landscape. In our everyday lives, even when most of us are completely unaware of it, our surroundings and the economic activities that go on in them are still affected by the glaciers that disappeared 10,000 years ago.

Conclusion

Glaciers have a significant impact on human activity, for both good and bad, both locally and globally. Glacier hazards are tragically familiar in glaciated areas, and glacier resources of various kinds have been exploited in the context of different technological perceptions and capabilities. Although technology has played a role in the mitigation of glacier hazards, hazard mitigation usually relies heavily on hazard prediction, which glaciologists are not yet able to provide at the level of sophistication required.

Glaciers have provided resources for a range of human activities since prehistoric times, and our ability to exploit glacial resources is likely to increase in the future as our knowledge and understanding of glaciers, and glacial landscapes, continues to grow.

CHAPTER 7

CONCLUSION

One of the most exciting things about glaciers is that although we know they play a really important role in the global environment, we still don't fully understand everything about them.

We have seen in this book that a great deal is known about how glaciers fit into the great global system: how they grow and shrink over time, move, alter landscapes and affect human activity. However, we have also seen that there are a lot of questions about glaciers to which science does not yet have the answers. How will glaciers respond to future climate change? How will glaciers affect future sea levels? Exactly where have glaciers occurred at different times in geological history? How, exactly, are certain landforms such as drumlins formed? Questions such as these are what future research into glaciers and glaciated environments needs to address.

Unanswered questions about glaciers span the whole breadth of the topic. For example, at a global scale we don't even know for sure whether the world's two ice sheets are currently growing or shrinking, and at the micro-scale of glacier physics we don't know exactly how tiny debris particles in ice influence how it flows. Uncertainties remain both about the mechanics of individual processes and about how processes interact to control large-scale glacier behaviour. Big questions such as how ice sheets will respond to climate change need to be broken down into sets of smaller questions about the properties of ice, water and sediment in different thermal and pressure environments. The answers to the small questions then need to be fitted together to answer the big questions. Tying together the answers to the small questions about individual parts of the glacier system and using them to answer big questions about whole ice sheets is the business of ice-sheet modelling, and this is one of the most important and active areas of modern research (Information Box 13).

Like all science, ice-sheet modelling depends on accurate information. Unfortunately, accurate information about glaciers is hard to obtain. For example, it is difficult to measure temperatures underneath a 3km-thick ice sheet! However, there have recently been some major advances in glaciological data collection related to satellite remote sensing. Remote sensing is another major area of progress that will carry glacier science forward in the future (see Information Box 14).

Information Box 13: Ice-sheet modelling

Ice-sheet modelling involves using what we know about how glaciers work to calculate mathematically how they would behave under different conditions. In this way we can:

- make predictions about future changes in glaciers
- work out how glaciers behaved in the past.

Most models involve dividing the glacier into very small sections, taking values for parameters such as temperature, snowfall and surface gradient, and using them to calculate values for other parameters such as velocity. Accurate modelling requires detailed understanding of the processes and parameters that control glacier behaviour, such as the amount of snowfall or the temperature at the bed, but modellers don't always have all the information that they need about the glaciers that they are modelling. However, making models based on simplifying assumptions that don't match reality is not always useful. For example, oversimplified models of ice sheets that only work for rigid, smooth, horizontal beds and do not allow for the occurrence of ice streams are not much use for modelling real ice sheets with complex beds and ice streams. The challenge for modellers in the twenty-first century is to model realistic glaciers and ice sheets. To achieve this, developments in modelling need to progress hand in hand with new observational glaciological data.

You can find many examples of ice-sheet modelling on the internet. For example, look at pages from the University of British Columbia Glaciology Group or the University of Maine Ice Sheet Model.

Information Box 14: Satellite remote sensing and GIS

Satellites in orbit around the Earth now gather huge amounts of information about glaciers. Remotely sensed data are now a major part of the study of glaciers and glacial landscapes. For example, in 2003 NASA launched its 'Ice, Cloud and Land Elevation Satellite' (ICESat) to provide data about ice-sheet growth and shrinkage. The study of landscapes also uses remotely sensed data. Patterns of drumlins and other landforms that cannot be seen clearly from ground level, for example, show up clearly on satellite images, and satellite data is a major tool in glacial geomorphology.

Remote sensing and geographical information systems (GIS) are probably the most important new developments in the study of glaciers since deep ice coring. Information about surface conditions and about the precise extent of glaciers is more abundant now that it has been in the past, and day-to-day monitoring of glaciers and glacial hazards is now routinely accomplished by satellite imagery. A good place to start exploring some of the imagery that is available is through NASA's Earth Observatory website (see e.g. Figure 33).

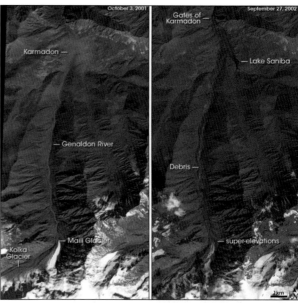

Figure 33: Satellite images indicating the extent of the collapse of the Kolka glacier, Mount Kazbe in southern Russia, on 20 September 2002. Images by: Robert Simmon based on ASTER data, NASA, Earth Observatory.

Glacier science is typical of many areas of geography in that unifies several different disciplines, and it considers together many different parts of the environmental system. In the study of glaciers and glacial landscapes, thermal regime, hydrology, substrate rheology (deformability of the glacier bed), ice deformation, mass balance, geomorphology, glacier dynamics, and the other topics that we have considered in this book all fit together in a single complex system. Glaciers play a key role in the global environment, both contributing to, and serving as indicators of, environmental changes. There are many 'Grand Unsolved Problems' in the study of glaciers, and in physical geography more broadly, and the continuing search for a unified theory that links the material properties and dynamic processes of ice, the morphology of glaciers, the features of the landscape and nature of environmental change is one of the many exciting prospects in the study of glaciers and glacial environments in the twenty-first century.

Activity Box 23: Scientists' most recent discoveries about glaciers

One way of learning how much we do or do not know about glaciers is to see what scientists are currently working on and what they have most recently discovered. New discoveries about glaciers are widely reported, because they are important for topics of general interest such as climate change and sea-level rise. Articles about new glacier science appear in national newspapers and on the websites listed at the end of this book, as well as on television and radio.

- Search through, watch or listen to a number of articles or features and list all the items that include science news about glaciers.

- Write down the new discovery or main point of each item. (You may wish also to think about who has written the article and identify any bias in reporting and possible reasons for that bias.)

- Identify on a map the location that the item refers to.

- Find a section in this book that covers a topic related to the item.

- Devise a new sentence or paragraph to up-date, or add this new example to, what this book says.

REFERENCES AND FURTHER READING

Books and journal articles

Addison, K. (1997) *Classic Glacial Landforms of Snowdonia* (second edition). Sheffield: Geographical Association.

Baker, V.R. and Bunker, R.C. (1985) 'Cataclysmic late Pleistocene flooding from glacial Lake Missoula: a review', *Quaternary Science Reviews*, 4, pp. 1–41.

Benn, D.I and Evans, D.J.A. (1998) *Glaciers and Glaciation*. London: Arnold.

Bennett, M.R. and Glasser, N.F. (1996) *Glacial Geology*. Chichester: John Wiley & Sons.

Boulton, G.S. and Jones, A.S. (1979) 'Stability of temperate ice saps and ice sheets resting on beds of deformable sediment', *Journal of Glaciology*, 24, pp. 29–42.

Dugmore, A.J. and Sugden, D.E. (1991) 'Do the anomalous fluctuations of Solheimajökull reflect ice-divide migration?', *Boreas*, 20, pp. 105–13.

EPICA community members (2004) 'Eight glacial cycles from an Antarctic ice core', *Nature*, 429, pp. 623–8.

Fleisher, P.J. and Sales, J.K. (1972) 'Laboratory models of glacier dynamics', *Geological Society of America Bulletin*, 8, pp. 905–10.

Glen, J.W. (1955) 'The creep of polycrystalline ice', *Proceedings of the Royal Society Series A*, 228, pp. 519–538.

Grove, J.M. (1988) *The Little Ice Age*. London: Methuen.

Hambrey, M. (1994) *Glacial Environments*. London: UCL Press.

Hammer, C.U. (1977) 'Past volcanism revealed by Greenland ice sheet impurities', *Nature*, 270, pp. 482–6.

Bond, G. and 13 others (1992) 'Evidence for massive discharges of icebergs into the north-Atlantic ocean during the last glacial period', *Nature*, 360, pp. 245–9.

Kamb, B. (1987) 'Glacier surge mechanism based on linked cavity configuration of the basal water conduit system', *Journal of Geophysical Research*, 92 (B9), pp. 9083-100.

Kamb, B. and LaChapelle, E. (1964) 'Direct observation of the mechanism of glacier sliding over bedrock', *Journal of Glaciology*, 5, pp. 159–72.

Kleman, J. and Borgström, I. (1994) 'Glacial landforms indicative of a partly frozen bed', *Journal of Glaciology*, 40, pp. 255–64.

Knight, P.G. (1997) 'The basal ice layer of glaciers and ice sheets', *Quaternary Science Reviews*, 16, pp. 1–19.

Knight, P.G. (1999) *Glaciers*. Cheltenham: Stanley Thornes.

Menzies, J. (ed) (1995) *Modern Glacial Environments: Processes, dynamics and sediments*. Oxford: Butterworth Heinemann.

Paterson, W.S.B. (1994) *The Physics of Glaciers* (third edition). Oxford: Pergamon.

Piotrowski, J.A. (2003) 'Earth science: glaciers at work', *Nature*, 424, pp. 737–8.

Richardson, D. and Reynolds, J.M. (2000) 'An overview of glacial hazards in the Himalayas', *Quaternary International*, 65-6, pp. 31-47.

Shaw, J., Kvill, D. and Rains, B. (1989) 'Drumlins and catastrophic subglacial floods', *Sedimentary Geology*, 62, pp. 177–202.

Souchez, R. and Lorrain, R. (1991) *Ice Composition and Glacier Dynamics*. Berlin: Springer.

Sugden, D.E. and John, B.S. (1976) *Glaciers and Landscape*. London: Edward Arnold.

Warren, C.R. (1991) 'Terminal environment, topographic control, and fluctuations of West Greenland glaciers', *Boreas*, 20, pp. 1–16.

Willis, I.C. (1995) 'Intra-annual variations in velocity motion: a review', *Progress in Physical Geography*, 19, pp. 61–106

The journal *Progress in Physical Geography* carries regular 'progress reports' on glaciers, and often features review articles on topics associated with glaciated landscapes. The journals *Nature* and *Science* carry the most up-to-date news about discoveries in glaciology.

Websites

Alaska Glacier Refreshment – http://www.alaskaglacier.com/

American Scientists Online – http://www.americanscientist.org/template/AssetDetail/assetid/26024

British Antarctic Survey – http://www.antarctica.ac.uk/About_Antarctica/index.html – All about Antarctica

BBC news – http://news.bbc.co.uk/science – Often carries news concerning glaciers and global environmental change.

Geomorphology from space – http://daac.gsfc.nasa.gov/www/geomorphology/ – A collection of satellite images and information including a section all about glaciers

Glacier – http://www.glacier.rice.edu/ – l about Antarctica and its ice sheet

Global Land Ice Measurements from Space – http://www.glims.org/

Holgate School – http://www.holgate.info/Environmental%20pages/EcoHolgate/fieldwork/Processes/glaciers.htm

Peter Knight's Glacier pages –
   http://www.petergknight.com/glaciers – Photographs,
   links and other resources as well as updated information
   for readers of this book.
National Geographic – http://news.nationalgeographic.com/
   news/2003/09/0923_030923_kilimanjaroglaciers.html
NASA – http://winds.jpl.nasa.gov/publications/
   greenland_fig_7.cfm
NASA Earth Observatory – http://Earthobservatory.nasa.gov/
   – An excellent resource of images, information and news
NASA Planetary Photojournal –
   http://photojournal.jpl.nasa.gov/catalog/PIA03899
Oxford Brookes University – http://ssl.brookes.ac.uk/
   Glacial_Environments_2003/page_13.htm
University of British Columbia Glaciology Group –
   http://www.geop.ubc.ca/Glaciology/
University of Innsbruck's Tropical Glaciology Group –
   http://geowww.uibk.ac.at/glacio/RESEARCH/
University of Maine Ice Sheet Model –
   http://www.ume.maine.edu/iceage/
US Geological Survey – http://pubs.usgs.gov/fs/fs133-99/ –
   Satellite image atlas of glaciers of the world
US National Climate Data Center for Paleoclimatology –
   http://www.ncdc.noaa.gov/paleo/icecore.html –
   Information about ice cores and climate change
US National Snow and Ice Data Center – http://nsidc.org/ –
   A wide range of educational resources about glaciers
Virtual Geomorphology – http://main.amu.edu.pl/
   ~sgp/gw/gw.html – A collection of articles covering a
   wide spectrum of geomorphic topics
World View of Global Warming –
   http://www.worldviewofglobalwarming.org/
   pages/glaciers.html

Note: Not all websites about glaciers are reliable, but
those operated by universities or by research
institutions such as NASA or the US Geological Survey
usually carry sound information. Beware of sites
specifically aimed at young children because they
often use out-of-date and oversimplified material; and
sites published by religious groups, which may not
base their information on reliable research.